Quadrants OF THE Corporeal

REFLECTIONS ON THE FOUNDATIONS OF EXPERIENCE

RANDOLPH LUNDBERG

ISBN: 978-1-4834-7221-8 (sc)
ISBN: 978-1-4834-7223-2 (hc)
ISBN: 978-1-4834-7222-5 (e)

Library of Congress Control Number: 2017911176

Lulu Publishing Services rev. date: 9/26/2017

Take me to that blessed Region where I in Thought shall see the insides of all solid things.

—Edwin A. Abbott, *Flatland: A Romance of Many Dimensions*

Contents

Introduction

This book lays out a view about how the conscious experience of human beings and other animals is related to the rest of what there is. The questions it considers are often discussed under the heading of "the mind-body problem" or "the hard problem of consciousness."

To the best of my knowledge, the view presented here is a novel one. How it resembles and how it differs from the views of various other writers will emerge in the course of the book. Many pages are devoted to these comparisons, including several whole chapters on comparisons that I consider especially interesting.

Views such as mine are often given names that end in "ism." I have no "ism" for my view, but I do have a name for it. Since one of its central characteristics is the delineation of a two-by-two grid of four mutually exclusive sets of properties, it can reasonably be called *the quadrants view*. I also have a name for an important component of the quadrants view: *the two-quadrant foundation hypothesis*.

After developing the quadrants view and convincing myself of its merits, I began to wonder why it doesn't already have advocates. There are many competing views of this subject that have long had advocates. Why not this one? One possibility is that the quadrants view involves an error that is apparent to everyone but me. The answer I favor, however, is that there are powerful forces at work that steer people's thinking in other directions. I will mention four such forces, which are at work in various combinations in almost everyone.

One force that has been highly influential historically, but is less so today, is belief in personal immortality. The belief that human beings continue living in an altered form long after their hearts stop beating is impossible to reconcile with the quadrants view.

A second force, whose influence has increased in modern times as belief in personal immortality has declined, is the notion that the continuing development of science and technology will give us the ability to fully understand the world. Uncritical veneration of science blocks the path to the quadrants view.

A third force, which contributes to the uncritical veneration of science and is also potent in its own right, is a widespread and deeply entrenched mind-set that I call *outwardism*. Human beings are naturally inclined toward outwardism, and outwardism is antithetical to the quadrants view. Details must wait until I lay some necessary groundwork.

Fourth and last is the disordered state of the terminology with which this subject is usually discussed. Some of the most commonly used words and phrases are so nebulous, so ambiguous, or so freighted with dubious presuppositions that it is all but impossible to use them without sinking into confusion. There are many such terms that I make a point of not using, for reasons that I explain as I go along. This book is not about words, but it contains many paragraphs that are about words—nuisance words that tend to derail the discussion of what this book is about. In the early stages of working on this project, I worried that avoiding so many widely used terms might make it impossible to write the book. Could it be that the terminology associated with this subject is so twisted that the truth cannot be told? Things are not quite that bad, I finally decided.

An aspect of the terminology problem that deserves special mention is the widespread habit of approaching the subject through pairs of contrasting terms. There is a virtual Noah's Ark of conventionalized verbal pairs—mind and body, mind and brain, mind and matter, mental and physical, subjective and objective, private and public, first-person and third-person. It is common for writers to take a stand for or against some form of "dualism," but the ironic fact is that the self-described nondualists are just as devoted to these verbal pairs as the self-described dualists are. Dualists and nondualists alike are verbal dichotomists—captives of a dichotomous terminology that blots out the fourfold taxonomy of the quadrants view.

The forces that steer people away from the quadrants view also have the potential to steel people against it. If you believe that individual human beings live forever, if you believe that scientific research can uncover all of nature's secrets, if you are a bastion of outwardism, or if you cannot let go of the entrenched dichotomous terminology, then you are likely to find yourself rejecting the quadrants view. But there is a flip side. If you question the ancient promise of immortality, if you are willing to reflect on the nature and limits of scientific discovery, if you can outwit your outwardism, and if you can break a verbal habit in response to an explanation of its ill effects, then you might come to appreciate the quadrants view as I do.

1

The Composition of Things

The atomists of ancient Greece and Rome maintained that everything is made of tiny atoms. Their conception of the composition of everyday objects is lovingly laid out by the Roman poet Lucretius in *On the Nature of Things*. The following passage depicts the interior of an atom as consisting of tightly packed and inseparable "points":

> To proceed: since in all these atoms which
> We can't perceive, there are points or vertices,
> Each point is irreducible, remains
> The minimal atomic part; it cannot
> Ever be isolated; it was and is
> The unitary base for something else.
> These points in phalanx, one after the other,
> In crowded order, constitute the atom.
> (Lucretius, Book One, 599–606)

The super-solid atoms were thought to be impermeable, indivisible, and immutable in every way. "Their tough walls outlast all blows," the poet boasts, as if advertising a miracle product (Lucretius, Book One, 486). The sturdy virtues of the atoms do not scale up to larger objects, however, because the atoms join together in a somewhat ramshackle way, forming structures that are shot through with regions of empty space. Such a structure, while not exactly a house of cards, has a way of falling apart.

According to this theory, the only respects in which one atom can differ from another are size, shape, and motion. Lucretius thought that the familiar properties of everyday objects were clues to the sizes, shapes, and motions of their constituent atoms. The following passage presents some typical inferences about size:

> It's simple, too, to figure out why lightning
> Can pierce to the quick with such a sharper force
> Than can the fire we light in earthly torches.
> For you may say that the sky-born flame of lightning
> Is subtler, made of tinier atoms which
> Allow it to pass through tighter crannies than
> This fire of ours can, risen from torch or cinder.
> Again, light passes through thin horn, while water
> Splashes against it. Why? Why else: light-atoms
> Are smaller than those of the clear life-giving water.
> (Lucretius, Book Two, 382–391)

Shape is a more complicated subject. Lucretius sees a variety of clues to the shapes of atoms. These clues include solidity and liquidity:

> And what feels hardened to us, tightly packed,
> Must be more deeply interlocked with atoms
> Hooked to each other, like a tangled thicket.
> Now in the first rank of such substances
> Stands adamant, long hardened in contempt
> Of all blows, and tough flint, and stiff-trunked iron,
> And brazen bars that screech shut in the locks.
> But out of smooth and roly-poly atoms
> Whatever flows like liquid must be made.
> For a swallow of poppy seed's as easy as water
> Since the rolling particles never glob together,
> But splash and spill like water when you strike them.
> (Lucretius, Book Two, 445–456) .

They also include the tastes of what people eat or drink:

> Here notice that a sip of milk or honey,
> Rolled on the tongue, will lend a pleasant feeling,
> While sickening wormwood and harsh centaury
> Pucker the mouth with the twists of their foul taste.

2

Easy for you to see that smooth and rolling
Atoms compose what touches the sense with pleasure,
While what seems harsh, bitter, these things are stitched
And held together in more hooked connections,
Which forces them to wedge and split their way
Into our senses, slashing through the body.
(Lucretius, Book Two, 399–408)

These passages speak to the shapes of individual atoms and also to the way these shapes determine how the atoms fit together.

One could say that the ancient atomists imagined a world of tiny Lego blocks. Shaped to interlock, these tiny blocks were thought to enter into temporary assemblies, thereby making all the objects that are large enough for us to see or touch. It's a shame that Lucretius had no illustrator. It would be fascinating to see some drawings that show super-solid atoms of assorted shapes and sizes gripping each other as Lucretius thought they did inside objects of various sorts.

<>

This ancient theory of atoms has been abandoned. Yet the word "atom" lives on. It has found a new home in the very different conception of the composition of things that is given to us by modern science. What we call an atom today has none of the simple virtues that Lucretius attributed to atoms in his poem. It is a complex system that consists of one or more electrons moving in complicated ways around a nucleus that consists of protons and neutrons. The electrons and protons are equal in number and of opposite electrical charge, making the atom as a whole electrically neutral. By convention, we say that electrons are negatively charged and protons are positively charged, but we could just as well say the reverse. The electrons and protons are attracted to each other by the electrical force, which keeps the electrons close to the nucleus under most circumstances. The protons and neutrons in a nucleus are attracted to each other by nuclear forces, which keep the nucleus intact under most circumstances.

We still say that everyday objects are made of atoms. However, this statement does not match up with the modern theory as well as it does with the ancient one. The modern theory is better expressed by saying that everyday objects are made of what atoms are made of—electrons and nuclei. Atoms are already complex and impermanent objects made of these smaller and much

3

more permanent components. In the world as imagined by Lucretius, if you zoom in on rocks or pieces of metal or the stuff on your dinner plate with a sufficiently high-powered microscope, you encounter solid Lego-like atoms that have exactly the same sizes and shapes that they have when they exist in isolation. The only difference is that the atoms that make up these objects are gripping each other in a Lego-like way so as to form a larger structure. According to modern science, on the other hand, if you zoom in on rocks or pieces of metal or the stuff on your dinner plate you do not, for the most part, encounter atoms at all. The reason is that electrons, at least those that are known as valence electrons, can move from one nucleus to another. Some atoms can lose one or more electrons, thereby becoming positively charged ions. Other atoms can gain one or more electrons, thereby becoming negatively charged ions. In any molecule that contains more than one nucleus, there are bonding electrons that are shared by two or more of the nuclei. When a chemical reaction occurs, some electrons give up their ties to nearby nuclei and form ties to other nearby nuclei. Aqueous solutions are typically characterized by so-called dynamic equilibrium, in which electrons and nuclei are continually changing their associations in such a way as to preserve the overall concentrations of the participating types of ions and molecules. In a conducting wire that carries an electric current, huge numbers of electrons stream past innumerable nuclei at high speed. Because such changes are always in progress, if you zoom in on a typical object what you encounter is not atoms but a kind of high-speed dance of electrons and nuclei. These are the invariant components that maintain their integrity as they enter and exit various temporary arrangements. According to modern science, the composition of everyday objects is electronuclear.

I like to think of the electronuclear composition of things as a kind of fabric. Instead of horizontal and vertical threads, there are electrons and nuclei. In a piece of cloth, each horizontal thread crosses many vertical threads and each vertical thread crosses many horizontal threads. Likewise, in an electronuclear object each electron can associate with many nuclei and each nucleus can associate with many electrons. Unlike the literal fabric of a shirt or a towel, however, the metaphorical electronuclear fabric is extremely dynamic, with electrons continually in motion around nuclei and not infrequently changing their nuclear partners. Metaphorically speaking, everyday objects are made of a dynamic electronuclear fabric.

The word "atom" is so commonly used that you might think we couldn't get along without it. But in general, one can give a more accurate description of the composition of things using other words—electron, nucleus, molecule, ion, crystal, and so on. For example, it is often said that molecules such as carbon monoxide (CO) and hydrochloric acid (HCl) are diatomic, but one can make the same point by saying that these molecules contain two nuclei. As another example, it is often said that the different chemical elements represent atoms of different types, but the distinctions among the chemical elements are at bottom distinctions among nuclei. For example, a one-proton nucleus is a hydrogen nucleus, while a six-proton nucleus is a carbon nucleus. As chemical elements are distinguished by the number of protons in a nucleus, so the isotopes of an element are distinguished by the number of neutrons that accompany the protons. For example, a nucleus that is simply a proton is ordinary hydrogen, while a nucleus that consists of one proton and one neutron is deuterium, a heavier isotope of hydrogen.

It would be going too far to say that atoms don't exist at all. The nuclei of the chemical elements known as the noble gases, including helium, neon, argon, and xenon, never join with other nuclei in multinuclear molecules, and they rarely gain or lose electrons. Thus, these chemical elements do exist primarily in the form of atoms—single nuclei surrounded by a number of electrons equal to the number of protons in the nucleus. But apart from the noble gases, today's atoms exist only as transient formations within dynamic electronuclear fabric. The persisting threads of the fabric are the electrons and the nuclei.

A striking feature of electronuclear fabric is the extreme contrast of mass density within it. Mass density is the ratio of mass to volume, so one has to consider both mass and volume to compare densities. Protons and neutrons are roughly 1800 times heavier than electrons. Therefore, in objects having roughly equal numbers of electrons, protons, and neutrons, only about 1 part in 3600 of the total mass—less than 0.03%—is in the electrons. The rest of the mass—more than 99.97%—is in the nuclei. Yet while the nuclei possess almost all of the mass, they occupy almost none of the volume. In a typical molecule, the distance between neighboring nuclei is about 10^5 (one hundred thousand) times the diameter of a nucleus. The volume ratio is the cube of this linear ratio, so molecular nuclei occupy only about 1 part in 10^{15} of the volume of a typical electronuclear object. Bringing these mass and volume figures together, one can compute the following:

+ The density of a nucleus is about 10^{15} times the density of an everyday object as a whole (almost all of the mass in about $1/10^{15}$ of the volume).
+ The average density between nuclei is about 1/3600 of the density of an everyday object as a whole (about 1/3600 of the mass in essentially all of the volume).

The density ratio of a molecular nucleus to the extra-nuclear region is therefore about 3600×10^{15}, or more than 10^{18}. The electronuclear fabric of everyday objects is characterized by this extreme density contrast.

With a similar set of computations, one can show that the average density of the extra-nuclear region of a typical solid object is about a quarter of the density of air. In this sense, even the densest objects on earth—blocks of granite or ingots of gold, for example—have an airy near-emptiness through most of their volume.

There is no familiar object whose density can be meaningfully compared to the density of a nucleus. But there is another kind of comparison that provides a perspective on nuclear density. If you could compress an everyday object so as to remove the extra-nuclear region completely and crowd all the nuclei together into a tight huddle, then each linear dimension would be reduced to 10^{-5} of its actual length. If you did this to the planet Earth, its diameter would be reduced to about 400 feet. So the density of a molecular nucleus is roughly equal to the density of a 400-foot-wide ball that contains the entire mass of the earth.

According to Lucretius, all atoms consist of "points in phalanx, one after the other, in crowded order." The modern picture of the composition of things does not encourage the idea that there is such homogeneity on the ground floor. The electron "threads" and the nuclear "threads" of electronuclear fabric have some very different characteristics.

People sometimes say that everyday objects such as rocks, pieces of metal, and foods are "physical objects" or "material objects." These are two of the terms that I do not use because of their ambiguity. I discuss the ambiguity of "material object" in chapter 8. I discuss the ambiguity of "physical object" and many other phrases that contain the adjective "physical" in chapter 9. Setting these terms aside, I say that everyday objects such as rocks, pieces of metal, and foods are electronuclear objects. By this I mean that they are made of the kind of electronuclear fabric that I have described in this chapter.

2

The Composition of Human Beings

The word "body" is sometimes used to refer to a dead body. Keith Campbell is not wrong when he says "your body … is what the undertakers bury when they bury you" (1984, 2). However, between a living human body and a dead human body there are huge differences, which Campbell's statement downplays. The mind-body problem is about living bodies, not corpses and cadavers. In this book, I always use the word "body" to mean a living body.

The word "body" is sometimes used in a sense that emphasizes and perhaps only includes what is below the neck. This seems to be the case, for example, when conversation turns to the muscle-bound body of a body builder or the curvaceous body of a beauty queen. In another familiar sense, however, your body extends from the soles of your feet to the top of your head. In this book, I always use the word "body" in this more inclusive sense. A living human body includes a head and the living brain inside it. It consists of everything that gets weighed when a clean, dry, naked person steps on a bathroom scale. One could make an exception for a prosthetic limb or the contents of the alimentary canal.

<>

A living human body is an electronuclear object—or, if you prefer, an electronuclear system. It sounds slightly off-key to call a living human body an object because the word "object" is normally reserved for simpler things. The important point is that living human bodies are like rocks, pieces of metal, and foods in being made of the kind of dynamic electronuclear fabric described in chapter 1.

A living human body has parts of many sorts and sizes. It contains organs, such as the brain, heart, and liver. It contains cells of many types, such as nerve

7

cells, muscle cells, and blood cells. It contains molecules of many types. A fact that can be startling the first time you hear it is that more than two thirds of the weight of a human body is water—H_2O molecules. Perhaps another 15% is protein molecules—long chains of amino acids that fold up in complex ways. Two other major types of biomolecule are lipids and nucleic acids. Molecules of all these types are continuously created and destroyed as electrons and nuclei change partners, but under normal circumstances the overall percentages remain about the same. The molecules are made of nuclei and electrons. More than 60% of the weight of a human body is in oxygen nuclei, most of which are in the water molecules. Roughly 20% is in carbon nuclei and roughly 10% is in hydrogen nuclei. The remaining 10% or so is divided among nuclei of many other types. The electrons, remember, are lightweight in the extreme; all of them together account for less than 0.03% of the weight of a human body, which amounts to about one ounce of a 200-pound man. The total number of molecular nuclei in an adult human body is more than 10^{27}—a billion times a billion times a billion. The total number of electrons is several times greater.

In describing human bodies, some writers speak of "levels." The following passage by Jaegwon Kim is typical:

> The bottom level is usually thought to consist of elementary particles ... As we go up the ladder, we successively encounter atoms, molecules, cells, larger living organisms, and so on. The ordering relation that generates the hierarchical structure is the mereological (part-whole) relation: entities belonging to a given level, except those at the very bottom, have an exhaustive decomposition, without remainder, into entities belonging to the lower levels. (Kim 1998, 15)

I think this is a mistake, or at least a misleading oversimplification. Where exactly are the boundaries that separate each level from the level immediately above it? Are all molecules from H_2O to the largest protein at the same level? Is a double-helix DNA molecule at a higher level than one of its helical strands? Are the mitochondria and other organelles in a cell at a higher level than the surrounding cellular material that is not part of any organelle? Is the gall bladder at the same level as the brain or at the same level as some part of the brain such as the thalamus or the amygdala? Such questions are misconceived and do not have correct answers.

Why do Kim and others endorse this idea? It seems to me that there is a "love of levels" built into the human intellect. We like to imagine structures that consist of levels. In antiquity, it was common to think of the world as a Great Chain of Being having levels that were occupied, in descending order, by God, angels, humans, animals, plants, and minerals. People design and build many things that have a structure of levels. Some examples are management hierarchies, classification systems, ranking systems, computer programs, career ladders, video games, and parking decks. However, no one has discovered a structure of levels in living things. This is a loose manner of speaking that reflects the speaker's love of levels but is not supported by careful observation.

I see no harm in speaking loosely of the levels of organization in a living thing—provided one does so in a sufficiently casual manner. The trouble is that writers often cross the line from casual talk to literal assertion. They advance propositions and arguments that make no sense unless there really is a definite hierarchy of levels in us. For example, Kim discusses various propositions that contain references to "level L." In doing so, he loses touch with the real composition of a living electronuclear system. There is nothing in us that can be meaningfully designated "level L," no matter what L stands for. A living human body has parts of many sizes, all intricately structured, but it does not consist of a hierarchy of levels. I make no argument that depends on this being true, but I avoid describing the human body as the level lovers do because I do not think these descriptions are correct.

Each of us has an electronuclear body. On this point there is general agreement. However, if you change the verb and say that each of us *is* an electronuclear body, controversy swirls. Some people hold what I call body-only views, according to which a living human being is exclusively a living human body, a complex electronuclear system. Others hold what I call body-plus views, according to which a living human being consists of a complex electronuclear system plus something else.

The quadrants view is a body-only view. So are most of the other views discussed in this book. I am interested mainly in disagreements that arise between people who hold body-only views. To justify this focus, however, I will explain why I find all body-plus views implausible. There are several reasons. First, body-plus views are difficult to make sense of if you go beyond vague generalities and try to describe them in detail. Second, there are strong arguments against all body-plus views. And third, all the arguments I know of that have been made for body-plus views have serious flaws. I will illustrate these points with two well-known body-plus views—substance dualism, represented by Descartes among others, and the "mental steam" view of T. H. Huxley.

<>

In one sense of the word, a substance is a kind of stuff, such as wood or water. In another sense, which is somewhat archaic, a substance is a durable, persisting object, such as a table or a ghost. Substance dualism involves both of these senses of the word. Its central claim is that a living human being is a union of two durable objects that are made, respectively, of two different kinds of stuff—a body object made of body stuff and a mind object made of mind stuff. Along with this claim goes the idea that the two objects are connected in a way that enables them to interact and function as a unit. For example, signals from the sense organs in the body object cross over into the mind object, and ideas for action in the mind object set the body object in motion. Although substance dualism dates back to an era when no one understood that a human body is an electronuclear system, it can be combined with this modern understanding of the body. The result is a form of substance dualism according to which a human being is a union of an electronuclear body and a mind object that is made of nonelectronuclear mind stuff.

My critique of substance dualism is in appendix A. If you consider substance dualism to be an attractive view, I suggest that you read appendix A before proceeding in order to make sure you are aware of the many points against it. If you feel confident that substance dualism is wrong, you can set a faster pace by skipping appendix A.

My name for T. H. Huxley's "mental steam" view is based on the steam-whistle analogy in the following frequently quoted passage:

> The consciousness of brutes would appear to be related to the mechanism of their body simply as a collateral product of its working, and to be completely without any power of modifying that working as the steam-whistle which accompanies the work of a locomotive engine is without influence upon its machinery. Their volition, if they have any, is an emotion indicative of physical changes, not a cause of such changes. (Huxley 1896b, 240)

After introducing this view in connection with nonhuman animals, Huxley goes on to say that he thinks it is equally true of human beings. His statement that consciousness is a "collateral product" of the body, combined with the steam-whistle analogy, gives a sketch of a body-plus view according to which the body, and in particular the brain, emits a nonbodily "mental steam" of which consciousness, volition, and emotion are aspects. The following frequently quoted passage also expresses this view:

> But what consciousness is, we know not; and how it is that any thing so remarkable as a state of consciousness comes about as a result of irritating nervous tissue, is just as unaccountable as the appearance of the Djin when Aladdin rubbed his lamp in the story. (Huxley and Youmans 1869, 178)

Although the only similarity that Huxley explicitly mentions here is being unaccountable, his selection of the image of the genie coming out of the lamp suggests that he held a similar conception of consciousness *coming out of* the brain. This kind of body-plus view is difficult to make sense of. The emission of genie-like "mental steam" by the body might bear a formal resemblance to the emission of hot, gaseous steam by a locomotive, but there would have to be huge differences. Hot, gaseous steam has an electronuclear composition, but "mental steam" does not. Hot, gaseous steam removes mass and energy from the locomotive that emits it, but "mental steam" does not remove mass or energy from the brain that emits it. How can we understand these requirements? Presumably Huxley did not think that the genie-like "mental steam" exits from the skull into the surrounding air; he must have thought of it as somehow hovering in the interstices of brain tissue. Granting this, how long does each puff of mental steam remain in existence after it is emitted? Can it undergo changes? By what process does it cease to exist? Huxley does not go into any of these details, and it is doubtful whether plausible details can be supplied. The sketchy description in the quoted passages seems to be a theoretical dead-end.

In making his case for the "mental steam" view, Huxley structures his argument as follows:

> For of two alternatives one must be true. Either consciousness is the function of a something distinct from the brain, which we call the soul, and a sensation is the mode in which this soul is affected by the motion of a part of the brain; or there is no

> soul, and a sensation is something generated by the mode of
> motion of a part of the brain. (Huxley 1896b, 210)

Huxley rejects the first alternative, which is substance dualism, and embraces the idea that consciousness is a "collateral product" of the brain as the only remaining alternative. The flaw in this reasoning is that it overlooks a third possibility: the properties that substance dualists attribute to a nonbodily soul and that Huxley attributes to a "collateral product" of the brain might be properties of the living electronuclear brain itself.

According to Huxley, the causation of a person's behavior runs through a brain that is devoid of consciousness, while simultaneously that brain gives off consciousness, sensation, and volition "as a collateral product of its working." Just as the steam that is emitted by a locomotive has no effect on what the locomotive does after emitting it, so consciousness, sensation, and volition have no effect on what a person's body does. This causal thesis later came to be called epiphenomenalism. Epiphenomenalism is usually discussed as a single doctrine, but I find it useful to distinguish two variants. Huxley's body-plus construal of consciousness and his causal isolation of consciousness go together in a natural way, yet they are logically independent of one another. If one can imagine body-plus mental steam at all, then one can imagine body-plus mental steam that can influence the brain that has just emitted it, thereby participating in the causation of bodily behavior. Conversely, if one can imagine body-plus epiphenomenalism as sketched by Huxley, then one can imagine body-only epiphenomenalism, according to which consciousness, sensation, and volition are properties of the living body that do not participate in the causation of bodily behavior.

I discuss epiphenomenalism in chapter 22, as part of a comprehensive discussion of the causation of human behavior. In this chapter I am concerned only with the body-plus aspect of Huxley's "mental steam" view. I see no credible way to add detail to Huxley's sketchy account of this view, and I see no reason to favor his account over the body-only view that consciousness, sensation, and volition are properties of a living electronuclear system. I therefore find the view that a person's electronuclear body emits "mental steam" extremely implausible.

Rejecting substance dualism for the reasons given in appendix A and rejecting Huxley's "mental steam" view for the reasons just given, I embrace a body-only view of human beings. A living human being is an electronuclear system that consists exclusively of the kind of dynamic electronuclear fabric that is described in chapter 1.

<>

The term "substance dualism" has a verbal cousin, "property dualism," which I discuss briefly to close this chapter. Although the term "property dualism" is widely used, it has no generally accepted meaning. There is general agreement that "property dualism" concerns a distinction between properties of two types, but this leaves two dimensions of ambiguity.

First, there is no consensus as to what the properties are properties *of*. Paul Churchland and Keith Maslin say "the brain":

> **Property Dualism.** The basic idea of the theories under this heading is that while there is no *substance* to be dealt with here beyond the physical brain, the brain has a special set of properties that are nonphysical: hence the term *property dualism*.... They are held to be nonphysical in the sense that they cannot ever be reduced to or explained solely in terms of the concepts of the familiar physical sciences. (Churchland 1988, 10; his emphasis)

> Property dualism eschews the notion of immaterial soul substances and maintains instead that the brain, a composite physical thing, possesses two fundamentally different sorts of property—mental features and consciousness in particular—that are non-physical in nature, as well as familiar physical properties such as mass, colour, and texture. (Maslin 2007, 31–32)

John Searle and David Chalmers say "the world":

> Property dualism is the view that there are two metaphysically distinct kinds of properties in the world, mental and physical. (Searle 1997, 168)

> The dualism implied here is instead a kind of *property dualism*: conscious experience involves properties of an individual that are not entailed by the physical properties of

13

that individual, although they may depend lawfully on those properties. Consciousness is a *feature* of the world over and above the physical features of the world. (Chalmers 1996, 125; his emphasis)

Dale Jacquette says "mind":

Property dualism maintains that mind has a dual nature, possessing both behavioral-material-functional and behaviorally-materially-functionally ineliminable and irreducible kinds of properties. (Jacquette 2009, 4)

Jaegwon Kim does not say what has the properties:

mental/physical property dualism—the view that mental properties are irreducible to physical properties (Kim 2005, 22)

It is not clear what all these different specifications of the property-bearer amount to, but they do not seem to be equivalent. For example, how should one classify Huxley's view that the brain has properties of one type and a "mental steam" emitted by the brain has properties of another type? This view is clearly not property dualism according to a "brain" definition (Churchland, Maslin) or a "mind" definition (Jacquette), but it might qualify as property dualism according to a "world" definition (Searle, Chalmers) or Kim's definition that does not specify a property-bearer.

Second, there is no consensus about the nature of the difference between the two property types. Property dualism is not just the innocuous claim that one can draw a distinction between two property types. It includes a further specification concerning how the two property types differ from one another. But what is the further specification? The Churchland, Kim, and Jacquette definitions specify that properties of one type must be "irreducible" to properties of the other type. Chalmers says that properties of one type must not be "entailed by" properties of the other type. Maslin says the two property types must be "fundamentally different." Searle says the two property types must be "metaphysically distinct." These criteria vary from the vague to the obscure. The classification of Searle's own view, which I discuss in chapter 13, is an interesting case in point. Searle's view features a dichotomy between two property types that he labels "subjective" and "objective," but he insists that he is not a property dualist. Other writers have objected to Searle's self-classification, sometimes intemperately. Maslin, for example, lets his exasperation show in the following passage:

> If Searle's position is not property dualism couched in other words, then I am afraid I do not know what it is. Despite his continual harping on about consciousness and intentionality being biological, natural phenomena—a very reasonable position in the light of the difficulties with full-fledged substance dualism—I believe that Searle should gracefully accept that he is, at the end of the day, a property dualist. (Maslin 2007, 164)

The reason for this disagreement, it seems to me, is the lack of a shared understanding of how two property types must differ if their co-existence is to count as property dualism. By Maslin's definition, property dualism involves two property types that are "fundamentally different." By Searle's definition, the two property types must be "metaphysically distinct." Evidently, Maslin considers Searle's two property types to be "fundamentally different," but Searle does not consider them to be "metaphysically distinct." Both writers could be right or—more likely—both writers could be making vague claims of indeterminate truth value.

Due to these ambiguities, use of the term "property dualism" breeds misunderstandings. Accordingly, it is best not to use it and to pay minimal attention to it when it is used by others. A writer who claims to be a property dualist might not be saying something wrong, but what exactly is he saying? To answer that question, one needs to ignore this ambiguous label and look at the details of the writer's claims and arguments.

<>

In summary, a human being is an electronuclear system. This body-only view is much more plausible than either substance dualism or Huxley's "mental steam" view. However, a human electronuclear system does not embody a literal hierarchy of "levels" and there is no illumination in either affirming or denying "property dualism." When describing a living human body, these notions should be avoided.

15

3

Outward Observation

Everything said so far about the electronuclear composition of things has been discovered by means of outward observation. "Outward observation" is my term for a way of observing things that involves spatial transmissions of one sort or another that travel from an observed object to an observer's sense organs. Dependence on spatial transmissions that travel from objects to sense organs is what makes outward observation outward. However, dependence on spatial transmissions does not suffice to define outward observation. There are several additional conditions, which I will explain after I give some examples.

The simplest cases of outward observation do not involve man-made technology. In unaided vision, an object emits or reflects light, which travels through space to light-sensitive eyes. In unaided hearing, a vibrating object generates compression waves, which travel through air or another medium to eardrums that are sensitive to the changing patterns of electric force that the arriving waves impress on them. In smelling, an object emits molecules, which diffuse through air to nasal linings that are sensitive to the changing patterns of electric force that the arriving molecules impress on them. In taste and touch, the spatially transmitted intermediary is just the electric force between the surface molecules of the observed object and the surface molecules of the observer's tongue or skin. These are the different types of spatial transmission that can depart from an electronuclear object and arrive at a human sense organ that is sensitive to them.

Man-made technology plays many different roles in outward observation. The variety and sophistication of the technology of outward observation is a testament to the importance of outward observation in our lives.

Let's start with some simple cases of technology-assisted vision. Instead of traveling directly from an observed object to an observer's eyes, light can pass through lenses or be reflected by mirrors along the way. The lenses in such instruments as eyeglasses, binoculars, optical microscopes, and optical telescopes improve and extend our powers of visual observation in many ways. Mirrors enable us to see things that we are not well-positioned to point our eyes at. This is a handy trick, as shown by the huge variety of mirrors. There are bathroom mirrors for looking at your face, automobile rear-view mirrors for observing the traffic behind your car, convex surveillance mirrors for monitoring the activity in large rooms, parabolic mirrors for collecting the light that enters a telescope, and mirrors mounted on bent sticks for looking at the hidden surfaces of teeth. The power of passing light from one mirror to another, as in a submarine's periscope, inspired Lucretius to write these appreciative lines some two thousand years ago:

> For whatever's stashed in the back rooms of a house,
> No matter how far, no matter how crooked the path,
> Can be teased out through a zigzag access of
> A series of mirrors, and appear anywhere.
> (Lucretius, Book Four, 302–305)

Cameras are vision-enhancing tools of a different sort. A camera captures light coming from one object and uses it to create another object, which we call an image. People can then learn about the original object by studying the image. With lenses and mirrors, the same photons travel all the way from the observed object to the observer. With a camera, photons travel from the observed object to the camera-made image; then, perhaps much later, different photons travel from the image to the observer.

If a spatial transmission from an observed object to an observer is likened to a trip, then any intermediate devices that process the transmission, such as lenses, mirrors, and cameras, divide the trip into legs. In any object-to-observer trip of two or more legs, only the final leg, which terminates at a sense organ, has to use a transmission medium that a sense organ is sensitive to. The other legs, and in particular the first leg, can use transmission media for which human beings have no sense organ.

There are many examples of object-to-observer trips whose first leg uses a transmission medium that we cannot sense. When we use X-ray telescopes to collect X-ray radiation from the sky, or make X-ray images of the body for medical purposes, or use X-ray diffraction to study the structure of molecules, the first leg of the trip uses X-rays. When we use radio telescopes to collect radio wave radiation from the sky, or radar to locate airplanes and bad weather, or magnetic resonance imaging (MRI) to study the body for medical purposes, the first leg of the trip uses radio waves. When we use infrared goggles to view a landscape at night, or infrared photography to make images of it, the first leg of the trip uses infrared radiation. In ultrasound imaging, the first leg of the trip uses high-frequency compression waves to which human ears are deaf. In sonar, the first leg of the trip uses compression waves in water. In a transmission electron microscope, the first leg of the trip is a stream of electrons that has passed through a target object. In a scanning electron microscope, the first leg of the trip can include both electrons and electromagnetic radiation of various wavelengths that are emitted by a target object when it is bombarded by a beam of electrons. In positron emission tomography (PET scans), the first leg of the trip begins with positrons emitted by a radioactive tracer. The positrons are annihilated in encounters with electrons, producing gamma radiation (electromagnetic radiation of an extremely high frequency) that travels to a detector.

With most of these technologies, the transmission that travels the first leg of the trip is used to construct a viewable image, which reflects or emits light to complete the final leg of the trip. However, it is also possible to produce numeric output using devices that measure the intensity of whatever is transmitted in the first leg. Numeric output is normally conveyed to the eyes using some sort of visual display, but it can also be conveyed to the ears with a speech synthesizer or to the fingertips using Braille.

A different class of examples involves the use of a probing object that is affected by the target object in some specific way. Probing objects that produce numeric output are called measuring instruments—thermometers, pressure gauges, scales, amp meters, and so on. A familiar example of a probing object that does not produce numeric output is a piece of litmus paper, which changes color when in contact with its target object. A less familiar example is a cloud chamber, in which fast-moving subatomic particles create visible trails that show the route they have traveled. We sometimes make use of natural probing objects in addition to man-made ones. Using the motion of trees to judge the speed of the wind is a familiar example. If a probing object is used to observe a target object, then the first leg of the trip is the process, whatever it may be, through which the target object affects the probing object. The design and manufacture of a measuring instrument or other probing object depends on implementing the particular process that constitutes the first leg.

<>

I turn now to the additional conditions that are part of my definition of outward observation. There are four of them. Each specifies a way in which a sensory experience that is dependent on a spatial transmission from an object to a sense organ can fail to be a case of outward observation.

First, sensory experience can serve purely as a source of pleasure, without being used to study an object that helped to cause it. In listening to music, for example, one typically simply enjoys the music, without using it to study the instruments, musicians, or electronic devices from which the music comes. A clever detective could use music as a means of outward observation, but that is not the usual role of music in our lives. Similarly, in smelling kitchen aromas, tasting a rich dessert, or feeling shower water massage your shoulders, you typically reap sensual pleasures with little or no concern for the objects from which the pleasure-inducing spatial transmissions come. Even vision, which is usually used to observe objects, can be a pure source of pleasure, as when one takes in a mountaintop vista or the fading hues of a summer evening. Such aesthetic or hedonic exercises of the senses are distinct from outward observation, which is marked by attention to an object from which the spatial transmissions come.

Second, you can focus your attention on your sensory experience itself, rather than on an object that plays a role in causing it. People attend to their sensory experience, for example, if it suddenly seems abnormal in some way, prompting them to worry that they might need to visit a doctor. One can also attend to normal-seeming sensory experience. Generally speaking, people do this less often, but it is an important pursuit for those with a professional interest in sensory experience, such as psychologists, philosophers, and artists. Attention to sensory experience itself is a case of *inward observation*, which is the subject of the next chapter.

Third, outward observation must be distinguished from using your sense organs to receive messages from other people. Spatial transmission to a sense organ is involved in both outward observation and interpersonal communication, but what sets interpersonal communication apart is the role of thinking at the source. In interpersonal communication, the spatial transmissions form spoken words, written words, diagrams, or signs of other types, which are used by a sender to express his thoughts. The recipient then uses the resulting sensory experience to understand the sender's thoughts. In outward observation, the spatial transmissions are not part of a system of interpersonal communication.

A subtlety here is that symbols used for interpersonal communication can be outwardly observed as objects. For example, one can study the shapes of a text font or the hands of a person who is using sign language, ignoring the associated message content. Thus, one cannot say for sure that a person looking at written text or the hands of a person who is using sign language is engaged in interpersonal communication and not outward observation. It depends on the context. If the person is using the text or hand signals as a path to the thoughts of the person who produced them, then that is interpersonal communication. If the person is studying the medium without any regard to the message, then that is outward observation. If the person is doing some of each, then he is multi-tasking, mixing interpersonal communication with outward observation.

Fourth and last, observation of the doings of other people belongs in a special category. Observation of human behavior includes visual observation of a person's motion, posture, and facial expressions. It can also include auditory observation of voice quality, hand-clapping, finger-snapping, or other man-made sounds. Observation of human behavior *involves* outward observation, but it is not *pure* outward observation, because the observer brings to bear, in addition to his visual or auditory capacities, a certain sense of human nature. When a tree moves in the wind we see only a commotion of branches, but when a person moves in certain ways we often "see" moods, intentions, and other human characteristics that one cannot literally see at all. When we say that someone looks tired, sounds upset, or seems to be in a hurry, we are verbalizing a judgment that includes this fellow-human factor. Pure outward observation does not include this fellow-human factor. Closely related to the observation of other people is the observation of nonhuman animals. Here too we sometimes "see" moods or intentions, in which case we are doing something more than pure outward observation.

A possible source of confusion here is the distinction between observations of human *behavior* and other sorts of observations of people. The examples of advanced medical imaging technology that I gave earlier illustrate pure outward observation of human tissue. Here the fellow-human factor does not mix in. The fellow-human factor is involved only when one is observing the behavior of a whole person, which one associates with certain invisible characteristics of the observed individual.

Outward observation as I have defined it here is related in a noteworthy way to the fields of study that we call the physical sciences. The data of the physical sciences are without exception obtained by means of pure outward observation. This is true because of the customary usage of the phrase "physical sciences."

Fields of study that combine outward observation with inward observation and/or the fellow-human factor are not classified as physical sciences. Some examples are psychology, economics, and linguistics. If we define the phrase "the sciences of pure outward observation" to include all and only those fields of study that are based entirely on the evidence of outward observation, then "the sciences of pure outward observation" and "the physical sciences" are co-extensive expressions, or very nearly so.

Outward observation is not limited to the physical sciences, however. There are different ways of using the word "science," but most people reserve this word for research projects that are marked by a certain degree of complexity. There are countless everyday uses of outward observation that are not part of any complex research project. For example, it is outward observation, but not what most people would call science, to notice that a door is open or that a coffee cup is empty. Outward observation is outward observation, whether it is part of a "scientific" research project or part of plain old everyday life.

<>

The claim that we observe objects by means of spatial transmissions that travel from the observed objects to our sense organs is almost universally accepted today. However, some famous thinkers have questioned it. I discuss four of these challenges, put forth respectively by Descartes, Berkeley, Kant, and Hume, in appendix B. If you have doubts that outward observation works as I have described it here, or if you are simply curious about these other views, you might want to read appendix B before proceeding.

<>

Much writing about "space" is bedeviled by the extreme ambiguity of this word. Central to this discussion of outward observation is the space that is occupied by electronuclear objects and traversed by the spatial transmissions to which our eyes, ears, and other senses organs are sensitive. Other important types of "space" include the spatial aspect of visual experience, which goes in and out of existence as you open and close your eyes, and a variety of mathematically defined spaces, among them Euclidean, elliptic, and hyperbolic.

Hoping to avoid both long-winded expressions and misunderstandings, I would like to have a conveniently short term that refers unambiguously to the space that is occupied by electronuclear objects. It is surprisingly difficult to find one. A term that is commonly used for this purpose is "physical space."

But "physical space" has a troublesome ambiguity, which is illustrated by the following passage:

> We conclude then that *mental* data have their own (phenomenal) kinds of spatiality; and that *physical* space is a theoretical construction introduced to explain the features and regularities of phenomenally spatial relations.... "Space" in the physical sense is an abstract theoretical ordering system. (Feigl 1967, 41–42; his emphasis)

Here, Herbert Feigl uses the term "physical space" to refer to a product of human thought, a variety of mathematical space that is believed to model the space that is occupied by electronuclear objects. He correctly notes that this theoretically constructed "physical space" is different from the "phenomenal" spatiality of visual experience. However, he does not note that both of these are different from the space that is occupied by electronuclear objects irrespective of all human experience and theory construction. For this space, Feigl provides no term at all. In *The Problems of Philosophy*, Bertrand Russell (2004b, 19) uses the phrase "the space of science" with the apparent intent of referring to the space that is occupied by electronuclear objects. One problem with this term is that several of the spaces under discussion here have important ties to science, so the phrase "the space of science" does not clearly distinguish them. The space that is occupied by electronuclear objects is studied by scientists, Feigl's theoretical space is constructed by scientists, and the space that is an aspect of visual experience is involved in all the visual observations that scientists make. All of these are spaces of science! A second problem is that calling the space that is occupied by electronuclear objects "the space of science" is misleadingly anthropocentric. The existence and nature of this space are independent of human scientific activity. The fact that human scientists study it is incidental. Max Velmans (2007, 352–354) contrasts "phenomenal space" with "real space," apparently using the latter term to refer to the space that is occupied by electronuclear objects. This usage misleadingly suggests that the spatial aspect of visual experience is not real. Every space under discussion here is real in its own way; it is their natures that are different.

My response to this terminological challenge is to give up the search for an unambiguous qualifier for the noun "space." From here on, I call the space that is occupied by electronuclear objects *the vast expanse.*

<>

Chapter 2 and appendix A make my case for a body-only view of human beings. They detail the difficulties that plague body-plus views such as substance dualism and Huxley's "mental steam" view. This chapter lays out the spatial-transmissions-from-objects account of outward observation, and appendix B details the difficulties that plague alternatives to this account. Combining a body-only view of human beings with the spatial-transmissions-from-objects account of outward observation yields the following sketch of the world. There is a vast expanse whose existence and nature depend in no way on human beings. Located here and there in this vast expanse are electronuclear objects. Some of these electronuclear objects are living human beings—us. From electronuclear objects emanate transmissions of various types—electromagnetic waves, compression waves, diffusing molecules, fields of electric force. We human beings have sense organs that enable us to use some of these transmissions to outwardly observe the electronuclear objects from which the transmissions come. If this sketch is correct, one can make the following three definitions.

First, an *outwardly observable property* of an electronuclear object is a property that some set of outward observations by some set of observers can produce evidence for. This is a broad definition, in two respects. First, it places no restrictions on the number or nature of the observers or the technology of observation. The observers do not have to be human observers or even observers of a type that currently exists, and the technology of observation need not currently exist either. Thus, the focus of the definition is the ability of an electronuclear object to affect the spatial transmissions that emanate from it. I assume that any way in which an electronuclear object can affect a spatial transmission can be used by a clever and well-equipped observer who receives the spatial transmission to learn about the electronuclear object at its source. Second, the definition does not specify any particular amount or strength of evidence. Any amount of evidence for the property is sufficient. Thus, there might be outwardly observable properties whose presence can be suspected, but not confidently believed in, on grounds of outward observation. An outwardly observable property must be real, however. If observers come to believe in the existence of a nonexistent property due to some combination of faulty observation and flawed reasoning, then that nonexistent property is not an outwardly observable property. Given the dependence of the physical sciences on outward observation, the set of outwardly observable properties includes all the actual and possible findings of the physical sciences.

Second, the *outward profile* of an electronuclear object is the set of all its outwardly observable properties. This set is larger than the phrase "outward profile" might suggest if taken out of context. It includes much more than the properties that a human being can discover with the unaided senses. It includes all the properties that any observer, human or not, and aided by any feasible technology, can find evidence for in any spatial transmission from the object.

Third, for any electronuclear object, one can draw a distinction between the object's outward profile and the *plenary object*. By the plenary object I mean the object with all of its properties, irrespective of how or whether the properties can be observed. Although the word "plenary" is not used very much, it seems to be the best word for this purpose. Other words with similar meanings, such as "whole," "full," "entire," and "complete," all carry a primary suggestion that no *part* of the object is excluded. It's true that a plenary object includes all of the object's parts, but that is not the focus of the contrast between a plenary object and its outward profile. The focus of this contrast is the different conditions that are imposed on *properties*. For a plenary object, the only condition is that a property must be a property of that object. For the outward profile of an object, the additional condition of being outwardly observable is imposed.

Using these definitions, one can pose the following question. How, in general, are plenary electronuclear objects related to their outward profiles? For example, what is the relation between a plenary acorn, a plenary water molecule, or a plenary living human body and its outward profile? There are two basic possibilities:

1. All the object's properties are outwardly observable properties. The outward profile of the object is the plenary object.
2. The object has one or more properties that are not outwardly observable. The plenary object is propertywise plumper than its outward profile.

For reasons to be given in coming chapters, the most plausible answer is that the second possibility is always the case; *every electronuclear object is propertywise plumper than its outward profile*. A related point, also developed in coming chapters, is that many people take it for granted that the outward profiles of most if not all electronuclear objects exhaust their plenary natures. In the minds of many, electronuclear objects have a purely outward reputation, which does not do justice to their plenary natures.

4

Inward Observation and Experience

I use the term "inward observation" in the way that many writers use the word "introspection." You inwardly observe a feature of your own experience when you notice it and pay attention to it while you have it. I prefer the term "inward observation" to "introspection" for two reasons. First, the parallel structure of the phrases "inward observation" and "outward observation" emphasizes the similarities between these two activities, which I spell out in chapter 5. Second, the word "introspection" has a certain ambiguity that I would like to avoid. It seems to me that in everyday usage the word "introspection" is more at home with thoughts and emotions than it is with sensations. Most people do not speak of hunger pangs or scalp itches as targets of introspection. It also seems to me that in everyday usage the word "introspection" can suggest the kind of inquiry that is sometimes called soul-searching, which goes beyond the mere noticing of features of experience. Why do I feel this way? Why did I do that? What really matters to me? Many would consider these introspective questions, yet you cannot answer them by simply noticing features of your experience. I need a term that covers the noticing of hunger pangs and scalp itches but not what people call soul-searching. I think "inward observation" meets this need better than "introspection" does.

In *The Principles of Psychology* William James (2007, chapter VII) uses the term "introspective observation" in a way that matches my use of "inward observation." U. T. Place (1956) uses the plural "introspective observations" in his article "Is Consciousness a Brain Process?" These terms avoid much of the ambiguity of "introspection." However, they contain a redundancy, because the "spect" in "introspective" signifies observation. If you remove the "spect" from "introspective" and turn what remains into a word, you get my term, "inward observation."

Bertrand Russell often uses the word "introspection," but in one book he opts for "self-observation":

> I say "self-observation" rather than "introspection," because the latter word has controversial associations that I wish to avoid. I mean by "self-observation" anything that a man can perceive about himself but that others, however situated, cannot perceive about him. (Russell 1995b, 99)

The definition of "self-observation" that Russell gives here makes his meaning clear, but without this special definition it is natural to understand "self-observation" more broadly as including a person's outward observation of himself. Actions such as looking at yourself in a mirror, weighing yourself, and inspecting your fingernails are examples of self-observation as normally understood. The term "inward observation" is more accurately focused.

<>

The word "experience" can be understood in several ways. I use it here in a broad sense that includes everything that anyone can inwardly observe, and more. After a brief survey of what people can inwardly observe, I note some ways in which experience extends beyond the limits of inwardly observable experience. I then compare my broad sense of "experience" with some other senses of this word.

A complete survey of what people can inwardly observe is impossible, for two reasons. First, there is too much variety, and second, the variety includes many subtleties that defy verbal description. Bertrand Russell once imagined the project of constructing a "minimum vocabulary for describing the world of my experience," in which "names are given to all the qualities of experiences" (1948, 265). A serious attempt to carry out this project would produce an ever-expanding dictionary much too large for anyone to master. Considering only color experience, for example, laboratory work shows that "human beings are capable of distinguishing something on the order of ten million colors" (Hardin 1988, 88). One could perhaps develop a system of alphanumeric codes to index ten million shades, hues, and blends, but a learnable vocabulary of that size and subtlety is out of the question. We do not have the requisite cognitive capacity.

Accordingly, the survey of inwardly observably features of experience that I present here is only a sketch.

It is usual to group features of experience into three broad categories—sensory experience, emotional experience, and thinking experience. Not all features of experience fall neatly into one of these categories, but this threefold division is nevertheless a useful way to organize the subject.

Much of our sensory experience belongs to what might be called the head senses. These are the four senses that depend on sense organs that are located exclusively in the head—vision, hearing, taste, and smell.

The features of human visual experience include colors, shapes, and sizes. A person's visual experience is, in part, an ever-shifting mosaic of colored regions, each having a shape and a size. The mosaic can contain hundreds or even thousands of such regions at a time, depending on the complexity of the current scene. However, if this were an adequate characterization of visual experience, all seeing would be like staring at abstract paintings. We see not just a mosaic of regions, but recognizable things—objects, happenings, relationships. The things are not part of visual experience, but visual experience includes visual representations of those things. When I look out my window, my visual experience includes visual representations of houses, trees, birds, people walking dogs, people driving automobiles, and other familiar elements of a suburban neighborhood. I recognize all these things in the mosaic of colored regions. My visual experience also includes a magnificent sense of three-dimensional visual space—not to be confused with the experience-independent vast expanse. My visual representations of objects are arrayed in my visual space. This visual content is ever-changing as new things happen in front of me or I look in different directions or the light changes.

The features of human auditory experience include all aspects of listening to music and speech. Musical experience includes the distinctive sounds of all the different musical instruments played singly or in combinations, as well as all the melodies ever composed. The sounds of people speaking have enough variety to represent all the words in every human language and all the unique voices of individuals. In addition to music and speech, human auditory experience includes an assortment of other sounds. There are sounds made by people—laughter, sneezing, coughing. There are sounds made by other animals—the barking of dogs, the singing of birds, the buzzing of bees. There are the sounds of automobile engines, rustling leaves, sirens, surf, tea kettles, and thunder. Much as we experience the visual representations of objects, we tend to experience sounds as heralds of their presumed sources. If I hear an unfamiliar sound that does not suggest any source to me, that fact immediately commands my attention and makes me wonder what the source of the sound is.

Most features of visual experience depend on light entering the eyes, and most features of auditory experience depend on sound waves entering the ears, but there are exceptions. If I close my eyes, I continue to have visual experience, of a sort that William James described as "a curdling play of obscurest luminosity" (2007, 620). On rare occasions, I see the bizarre glittering crescent of a so-called optical migraine. I have a continual soft hissing in my right ear—tinnitus. Covering my ears with my hands makes the hissing more noticeable, not less.

Human taste experience includes a staggering variety of flavors. One can easily fail to appreciate the extent of this variety because one normally tastes just one thing at a time, forgetting each flavor as the next one replaces it. One way to get a sense of the variety is to focus on a single letter of the alphabet. Consider, for example, the diversity of taste sensations produced in us by the following "c" foods: cabbage, cabernet, camembert cheese, cantaloupe, caramel, caraway seeds, carrots, cashews, cauliflower, caviar, cayenne pepper, celery, champagne, chardonnay, cheddar cheese, cherries, chicken prepared a hundred different ways, chickpeas, chili, chives, chocolate, cilantro, cinnamon, cloves, club soda, cocoa, coconut, coffee, cognac, cola, corn, cottage cheese, cotton candy, crabmeat, cranberries, cream, cream cheese, cream soda, cucumbers, currants, curry. The point can be made with other letters. Try it—it's an entertaining exercise.

Human smell experience is another grab bag. There are the odors and aromas of all the things that we can taste as well as many things that we either can't taste or choose not to. We can smell flowers, smoke, perfumes, sewage, air fresheners, bad breath, scented candles, skunks, paint.

It's traditional to identify five senses—the four head senses plus touch. But I find it difficult to draw a line between touch sensations and various other sensations that are not comfortably classified as touch. Here are some inwardly observable features of human sensory experience that do not belong to the head senses:

- The entire woeful variety of painful sensations—pulled muscles, broken bones, headaches, earaches, cuts and scrapes, intestinal cramps, pimples, sunburn, cracked lips, bee stings, sore throats, heartburn.
- Painless sensations that are due to bodily contact with objects—the feeling of wind on your face, your feet on the floor, your rear on a chair, a comb in your hair, your fingertips on a keyboard.

- Thermal sensations—gradations of hot, warm, cool, and cold.
- Sensations that accompany specific bodily functions, such as breathing deeply, chewing, swallowing, burping, sneezing, coughing, shivering, having a full bladder, having an orgasm.
- Miscellaneous sensations that are felt to have a specific location—itches, tickles, sticky or slimy feelings, hunger, thirst, pressure changes in the ears, dry eyes, pins and needles, nausea, numbness, the stuffiness of colds and allergies, electric shocks.
- Miscellaneous sensations that have a central or global quality—lightheadedness, dizziness, fatigue, drowsiness, fever and chills, inebriation.

That's enough on sensations; my aim is to give a sense of the variety, not to exhaust it.

Features of emotional experience are normally aroused by situations of various types in which we find ourselves. Among the common causes of emotional experience are romantic relationships, suspenseful competitions, disagreements, successes and setbacks, dilemmas, financial hardship, sickness, death, danger, unmasked deceptions, kindness, cruelty, and perceived injustice. There is one set of feelings that comes with being personally enmeshed in such situations and a different set of feelings that comes with viewing or reflecting on such situations that you are not part of. Portraying such situations in a way that evokes the associated spectator feelings is the stock in trade of novels, plays, and movies. Another source of emotional experience, which seems to engender some distinctive feelings all its own, is listening to music and songs. For example, many songs engender a pleasant sadness, which is different from the kind of sadness that you can't get rid of soon enough. Many words are associated with specific features of emotional experience—amorous, amused, angry, annoyed, anxious, ashamed, awe-struck, and so on through the alphabet. An impressive set of emotion-related words and phrases has been built from the word "heart": heart-warming, heart-breaking, heart-stopping, heart-wrenching, heartache, heartening and disheartening, heartfelt, faint-hearted, lighthearted, heartthrob, make your heart sink, tug at your heart strings, take heart, lose heart, have a heavy heart.

Features of thinking experience differ from one another less dramatically than do features of sensory and emotional experience, but the differences are real nonetheless. Here are a few types of inwardly observable thinking experience:

- An idea or a question suddenly occurs to you.
- You read or hear about something that reminds you of something else.
- Something you read or hear strikes you in a certain way—as false, dubious, well-put, unclear, hackneyed, interesting, funny, and so on.
- You have something "on the tip of your tongue."
- You daydream.
- You weigh the pros and cons of a decision.
- You lose your train of thought.
- You suddenly remember that you forgot to do something.

It seems that thinking experience of some kind or other is an almost continuous aspect of waking human life. If I am awake, then I am either engaged in some thought-intensive activity such as reading or writing or conversing, or I am engaged in some more or less thought-free activity such as driving a car or watering plants or playing tennis. If I am engaged in a thought-intensive activity, then of course I am having ongoing thinking experience. If I am engaged in some more or less thought-free activity, then I am typically also having ongoing thinking experience, because I am thinking casually about some unrelated topic—perhaps something I recently read or saw on television. Somewhere in my head there is a thought engine that rarely takes a pause.

There is another way of studying one's own experience in addition to inward observation: recollection. I can recollect many features of my past experience that I may or may not have inwardly observed while I was having them. I can recollect particular one-time features of experience, such as details of my wedding or the births of my sons. I can also recollect features of experience that I have had on many occasions, without being able to distinguish the occasions—the faces and voices of people I know, for example. The relation between an experience of recollecting and the past experience that is being recollected is a tricky subject. In general, the experience of recollecting is less detailed. It is a distillation of the original experience, and perhaps also a distortion. Experiences of recollecting can themselves be inwardly observed. However, they do not fit neatly into the threefold division of sensation, emotion, and thought. This is because the activity of recollecting has the character of thought, but the content of the recollection typically has a sensory and/or emotional character that reflects the nature of the past experience that is being recollected.

<>

We move on now from inwardly observable experience to experience that is not inwardly observable.

Sleep gives one a vacation from experience. It's a vacation with interruptions, however, because experience comes to sleepers in the form of periodic dreams. As far as I can tell, my dream experience is completely sheltered from inward observation. My dreams are typically experientially complex, with thoughts, purposive actions, visual scenery, emotions, conversations, and familiar people and places combining in a way that bears an uncanny resemblance to an episode of my waking life. But one of the differences between my experientially complex dreams and my experientially complex waking life is that while I am dreaming my capacity for inward observation is shut down. I am completely absorbed in the dreaming, unable to get an observational perspective on it. I can think and write and talk about my dreams later on only because I am able to recollect bits of them upon waking up.

Reading what others have said about dreaming and discussing dreaming with my friends have convinced me that for some people dreaming is not like this. For example, C. D. Broad writes:

> It is certain that many dream-experiences could have been introspected by the dreamer while they were happening; for I have quite often introspected my dream-experiences while dreaming, and I do not suppose that this is at all exceptional in people who are given to introspection. (Broad 1925, 373)

I find nothing to match this in my life, but for all I know Broad's way of dreaming is more typical than mine. As another example, much has been written in recent years about so-called lucid dreaming, which one writer describes as "a state in which a dreamer becomes aware that he's dreaming while the dream is still in progress" (Rock 2004, 150). Reportedly, there are lucid dreamers who can not only study their dreams as they have them, but also deliberately make things happen in their dreams and even send signals while dreaming by means of prearranged eye movements to waking people in a dream laboratory. This is all alien to me. My dreams can be vivid and gripping, but they are never "lucid" in this striking way. My cautious conclusion is that for at least some of us dream experience is less open to inward observation than waking experience is.

Inward observation and recollection have their respective advantages and disadvantages as ways of studying experience. The big advantage of recollection is that it provides access to certain realms of experience that are not accessible to inward observation. This includes some dream experience and all past experience. Past experience might have been inwardly observable while it was in progress but it will never be inwardly observable again. The big advantage of inward observation is that it provides access to more detail. There are many details of experience that are rarely or never retained in memory, which makes it impossible to study them through recollection.

There is good reason to believe that experience that is not inwardly observable includes most of the experience of nonhuman animals. If any animal with eyes has visual experience, then experience has been around for hundreds of millions of years and is today present in thousands of animal species. But there is little reason to think that the capacity to inwardly observe experience extends much beyond human beings. Can a dog inwardly observe its visual experience, thoughts, or emotional feelings? There is room for debate here, but I incline to a negative answer. When a dog yelps or scratches itself, it does seem that it must be inwardly observing a painful or itchy sensation, but even in these cases it could be that the animal is beset by these sensations without being able to focus attention on them in the way that people do. Animal expert Temple Grandin claims that in most cases nonhuman animals "act as if an injury or disease hurts them less than the exact same injury or disease would hurt a person" (2005, 185). If this is true, a plausible explanation is that nonhuman animals are less able to focus their attention on unpleasant features of their experience. I doubt that there is a clean line between animal experience that is and is not inwardly observable. Most likely, the human capacity for inward observation evolved by degrees, creating a gray area of experience that is "sort of" inwardly observable by the animal that has it. But the evidence suggests that the development of this capacity began relatively recently on the evolutionary time line, after the introduction of many animals that have experience but no capacity at all to inwardly observe it.

In addition to at least some human dream experience and much of the experience of nonhuman animals, there may be features of the waking experience of human beings that are not inwardly observable. It is tempting to suppose that while you are awake you can direct your attention to any detail of your current experience, much as you can point your eyes in any direction. But there are several reasons to think this is wrong.

First, many features of experience are so fleeting that they come and go in less time than it takes to shift your attention to them. If you try to focus your attention on something so ephemeral, you end up with only a recollection of what you had hoped to observe.

Second, many features of experience are effectively terminated by your attending to them because their existence depends on your attending to something else. If you are immersed in music and then try to study the details of your auditory and emotional experience, you break the spell that you were in. If you are thinking hard about a math problem and then try to study the details of your thought process, you derail that thought process, whose basis was a single-minded focus on the math problem. If you are angry or anxious about something and then try to study the details of your angry or anxious feelings, you take your mind off the reason for those feelings, which tends to make the feelings fade away. In all these cases, you get perhaps an inward glimpse of some departing feature of your experience, followed by a less detailed recollection of the departed feature.

Third, making any feature of your experience the target of your attention tends to change it slightly, if only to make it more distinct and vivid. You have a visual field with many colored regions and visual representations of many objects. If you now focus on one of these, what you focus on looks subtly different from the way it looked when it was a peripheral part of the scene. In cases where the change is small, you will be inwardly observing very nearly the same feature that existed before you began to observe it. Nevertheless, you are not inwardly observing exactly the same feature. If this is true, then the only fully inwardly observable features of experience are features of experience that are actually being inwardly observed. Features of experience that you have without inwardly observing them may be very similar to inwardly observed features of experience, but they are not exactly the same. In the midst of remarks similar to these, William James waxes metaphorical, lamenting that "The attempt at introspective analysis in these cases is in fact like seizing a spinning top to catch its motion, or trying to turn up the gas quickly enough to see how the darkness looks" (2007, 244).

In summary, there are at least three large categories of experience that are not inwardly observable:

+ Many features of human dream experience.
+ Many features of the experience of nonhuman animals.
+ Features of the waking experience of human beings that are too short-lived to observe or that are disturbed or transformed by having attention focused on them.

In these respects and perhaps others, the domain of all features of experience is larger than the domain of all inwardly observable features of experience.

<>

Writers often overlook or even tacitly deny the distinction between observable experience and unobservable experience. An interesting case in point is Thomas Nagel. In his landmark essay "What is it like to be a bat?" Nagel says that each animal species whose members have experience embodies a distinctive species-specific "point of view." He uses bats as his main nonhuman example because the bat "point of view" is presumably very different from the human "point of view." Bats rely heavily on a sonar-like echolocation system, which has no counterpart in humans. They divide their time between aerial hunting trips and hanging upside down in dark places—not common pastimes for human beings. So far, so good; I find it easy to agree with Nagel that the bat "point of view" is different from the human "point of view."

But Nagel's comparison of the two points of view goes astray in an important way. He focuses exclusively on differences in sensory experience, and says nothing about differences in the capacity for inward observation. He even writes as if the capacity for inward observation were the constant accompaniment of experience. For example, he makes the following comments about experience in general:

> ... the facts of experience—facts about what it is like *for* the experiencing organism—are accessible only from one point of view ... (Nagel 1979c, 172; his emphasis)

> ... the subjective character of experience is fully comprehensible only from one point of view ... (Nagel 1979c, 174)

What he should say here is that *some* experiential properties, especially human ones, are accessible and fully comprehensible from only one point of view, while other experiential properties, including most that belong to nonhuman animals, are not accessible from any point of view. Bats have a distinctive point of view, but it is a point of view on their environment and not on their own experience. Their experience is part of what constitutes their point of view, but it is not something that they have a point of view *on*.

According to Nagel, it is futile to try "to understand the experience of another species without taking up *its* point of view" (1979c, 172; his emphasis). This statement may be literally true, but it wrongly suggests that members of that other species are in a better position to understand their experience than humans are. In general, they are not. Granted, we human beings cannot understand bat experience because we do not have bat experience. But bats cannot understand bat experience either, for a different reason. They have the bat experience that human beings lack, but they lack the capacity for inwardly observing experience that human beings have. This is the point that Nagel's essay obscures and even seems to tacitly deny. Nagel seems to be imagining a kind of psychologically hybrid bat-human in which the sensory experience of a bat is combined with the human ability to inwardly observe experience. This is an interesting idea, but it is not a description of any real animal.

It is possible that I am being unfair to bats. Maybe they have a greater capacity for inward observation than I am giving them credit for. But bats are not the issue. The important point is that there is no necessary connection between having experience and having the capacity to inwardly observe experience. In the course of evolution, experience came first, setting the stage for the inward observation of experience, which came later and is less widespread. Even in human beings, there are many instances of features of experience that are unobservable or marginally observable—either because they are too short-lived or too dependent on a person's attention being directed elsewhere or part of a dream in a person who dreams as I do.

Commenting on Nagel's essay, Robert Van Gulick refers to cases in which "the system being understood and the one doing the understanding are one and the same, as they are in consciously self-understanding systems, such as a human or a bat" (2004, 389). Here the denial of a difference in the ability to inwardly observe experience appears in an explicit form. I doubt that bats are consciously self-understanding systems except in some extremely diluted sense. Even if I am wrong about bats, it is almost certainly the case that there are many animal species on this planet whose members have experience of one sort or another but cannot reasonably be called consciously self-understanding systems. On the time line of evolutionary novelty, there was a period during which some organisms developed an experiential point of view on their environment, and there was a later period during which some organisms with an experiential point of view on their environment developed an inward point of view on their experience. Experience does not in general come with a capacity for inward observation, let alone self-understanding.

In another essay, Nagel writes:

> Phenomenological facts have to be in principle, though not infallibly, introspectively accessible. (Nagel 2002, 213)

It would be unwise to disagree with this statement outright because it is open to different interpretations. What does Nagel mean by "facts"? How much of a hedge is implicit in the phrase "in principle"? However, this statement plainly suggests that where there is experience there is introspective access, and that is not the case. In general, on this planet, where there is experience there is little or no introspective access. Human beings, with their wide-ranging introspective access to their experience, are the exception rather than the rule.

<>

Inward observation of current experience and recollection of past experience fall under a broader category of cognition that I call inward access. You also have inward access to your beliefs, opinions, preferences, and vocabulary. You access beliefs, opinions, and preferences when you express them verbally or merely reflect on them. You access words that are in your vocabulary whenever you use them in conversation, in writing, or in thought. What characterizes each of these cases is an avenue of access that each person has to certain aspects of himself but not to anything outside himself. It's true that with the help of interpersonal communication one person can gain a kind of access to another person's experience, memories, beliefs, opinions, preferences, and vocabulary, but only the owner can access such things inwardly, without depending on interpersonal communication.

All these forms of inward access play important roles in our lives. To function in almost any job, you need inward access to beliefs and vocabulary that are essential qualifications for the job. One purpose of school examinations is to find out whether students can access certain items inwardly. When something finds its way onto a student's exam paper that the student did not access inwardly, we call that cheating. The delivery of many professional services, such as those of doctors, lawyers, psychological counselors, and financial consultants, depends on the client's ability to inwardly access relevant sensations, emotional feelings, and thoughts. The to-and-fro of casual conversation exercises the capacity for inward access of each participant.

<>

The broad sense of "experience" that I have been delineating contrasts with narrower senses of the word. I will mention two narrower senses to make it clear that I am not using the word in these narrower ways.

First, the word "experience" is often used in a way that includes a presumption of resultant learning. This is the sense that is found in Help Wanted ads. If an employer asks for five years of experience in a certain area, he does so because he believes those five years of experience will have had a cumulative effect on the job applicant that makes him better able to do the job. This is a reasonable belief because people do tend to "learn from experience," but it need not be correct in every case. A person who has no ability to learn, perhaps due to some disastrous disorder of the nervous system, might nevertheless lead an experiential life in the broad sense that he has an ongoing stream of sensations, emotional feelings, and thoughts during his waking hours. But he would not have experience in the sense that matters to an employer. His experience would have come and gone without producing any lasting change in him, much as a movie comes and goes without modifying the movie screen.

Second, a different narrower sense of "experience" is implicit in the customary explanation of the distinction between *a priori* knowledge and *a posteriori* knowledge. It is often said that *a posteriori* knowledge "depends on experience" whereas *a priori* knowledge "does not depend on experience." However, typical examples of *a priori* knowledge do depend on experience in the broad sense. Consider "knowledge of conceptual truths" as described in the following passage:

> Knowledge of conceptual truths can be obtained from reflection on the concepts involved, and need not rest on experience.... Knowledge obtained in this way is a priori knowledge. (Ludwig 2003, 8)

Since reflecting on concepts involves thinking experience, knowledge that depends on such reflection depends on experience in the broad sense. A second example is *a priori* knowledge of a theorem in mathematics, which depends on the features of thinking experience that constitute a personal trip through a proof of the theorem. Evidently, writers who explain the distinction between *a priori* knowledge and *a posteriori* knowledge in terms of dependence on experience are using the word "experience" in a narrower sense that allows them to say that a person can reflect on concepts or master a theorem in pure mathematics independently of experience.

I do not believe I have ever seen an explanation of the *a priori/a posteriori* distinction in terms of dependence on experience that includes a discussion of what counts as "experience" in the relevant sense. Thus, I am not sure how well defined this second narrower sense of "experience" is. One might guess from the examples just given that it excludes thinking experience, but that seems too simple. Suppose a mathematician working on a difficult proof keeps a log of his thinking about the problem, including such things as changes in his approach, computations that he carries out, and flashes of insight that he has. Successful at last, he publishes two documents—a presentation of his proof and a record of his thinking about the problem. I believe that most writers who use the terminology of *a priori* versus *a posteriori* knowledge would say that the presentation of the proof expresses the mathematician's *a priori* knowledge of mathematics whereas the log expresses his *a posteriori* knowledge of some of his recent thoughts. Yet both documents seem to be based on thinking experience, and indeed on the very same thinking experience. Regardless of how this riddle might be solved, there is a narrower sense of experience at work in the traditional distinction between *a priori* and *a posteriori* knowledge.

<>

Writers who discuss experience often use other words that have closely related meanings. I make minimal use of these other words. As much as possible, I discuss experience using the word "experience." However, these other words inevitably come into play when I discuss the work of the writers who use them. On those occasions, it is important to understand how a writer's use of one of these other words is related to experience in the broad sense. I conclude this chapter with brief introductions to some of these related words.

Conscious and consciousness. These words are sometimes used in a broad sense that is roughly coextensive with the broad sense of "experience." An animal is conscious, in this broad sense, if and only if it is having some experience. Whenever an animal is awake or dreaming, it is conscious in this broad sense and it is having some experience. When an animal is neither awake nor dreaming, it has no experience and it is not conscious in this broad sense.

"Conscious" and "consciousness" are also used in a variety of narrower senses. For example, they are sometimes limited to what a person's attention is focused on. Consider a person involved in a conversation whose attention is focused on the other person's spoken words and their meaning. His consciousness, in the attention-dependent sense, includes his auditory experience of the speech and the thoughts they trigger, but it does not include his concurrent visual experience of nearby objects, which has a background character. His consciousness in the broad sense includes his background visual experience in addition to the stream of speech he is attending to.

Some writers use the phrase "conscious experience." If "conscious" is used in the broad sense, this phrase is redundant and serves only to add emphasis; all experience is conscious experience. If "conscious" is used in the attention-dependent sense, this phrase means experience that is at the focus of attention. In this case, much experience, including most experience of nonhuman animals, is not conscious experience.

My subject is experience. Statements about consciousness obviously bear on my subject, but exactly how they bear on it can vary from one writer to another. Here are some examples from coming chapters. John Searle and Colin McGinn both seem to use the word "consciousness" in the broad sense; this makes their statements about consciousness directly comparable with my statements about experience. Gregg Rosenberg makes it clear that he uses "consciousness" more narrowly than "experience"; it is his statements about experience that are directly comparable with my statements about experience. Gerald Edelman writes about "consciousness" in a way that makes it unclear whether he is discussing experience in general or only certain more highly evolved forms of experience. This forces me to discuss his work under two different interpretations.

Even when the word "consciousness" is used in the broad sense that is roughly coextensive with the broad sense of "experience," it tends not to suggest the huge variety of different features of experience that I sketched earlier in this chapter. Consciousness in the broad sense is a single property, the property that all these features of experience have in common. This is the main reason why I prefer the word "experience": it makes it harder to lose sight of the full complexity of the subject. Some writers tackle the subject of "consciousness" as if there were just one thing in need of explanation, but in fact there are a great many things—the innumerable features of sensory, emotional, and thinking experience, in human beings as well as other animals.

Phenomenal. This adjective is often used with a meaning similar to "experiential." However, it seems to be used mainly for sensory experience, and less often for emotional experience or thinking experience. A writer who makes statements about "phenomenal reality" or "phenomenal consciousness" is discussing features of experience, but perhaps not all of them.

Quale and qualia. These nouns, singular and plural respectively, have been given a variety of meanings by academic writers but may be unfamiliar to the general reader. They are often used to mean something like "features of experience," but, like the adjective "phenomenal," they are used most often for sensory experience. Moreover, they are typically focused even more narrowly on very simple features of sensory experience such as an itch or a sensation of red. Features of thinking experience, features of emotional experience, and even complex features of sensory experience such as the flickering of a flame, a musical melody, or the visual recognition of a face, are not normally used as examples of qualia. A writer who discusses "qualia" is discussing features of experience, but perhaps not all of them.

Mind and mental. David Chalmers contrasts what he calls "two concepts of mind." One is experience-based:

> This is the concept of mind as conscious experience, and of a mental state as a consciously experienced mental state. (Chalmers 1996, 11)

The other is behavior-based:

> This is the concept of mind as the causal or explanatory basis of behavior. A state is mental in this sense if it plays the right sort of causal role in the production of behavior, or at least plays an appropriate role in the explanation of behavior. (Chalmers 1996, 11)

Although many writers use "mind" and "mental" in one of these two ways, it seems to me that neither of these definitions captures the everyday usage of these words. I think most people would say that anything you can access inwardly is mental whether or not it is experiential and whether or not it has any causal connection to your behavior. Your memories, beliefs, opinions, preferences, and vocabulary, all of which are stored in a nonexperiential form, are in this sense "mental" and aspects of your "mind." Various characteristics of a human being, such as those associated with words like "clever," "scatter-brained," and "methodical," are also commonly considered "mental" and aspects of the "mind." I am not proposing a third definition, but merely noting that the definitions Chalmers gives do not line up with ordinary usage. The words "mind" and "mental" are commonly used in a broad and nebulous manner that includes features of experience but also includes a good deal more. John Searle has it about right when he says "The notion of a mind is somewhat confused and lamentable" (1998, 40). A writer who uses these words might or might not be discussing experience, and might or might not be discussing other things in addition to or instead of experience. To determine how a writer's statements about "the mind" or "mental properties" relate to experience, one has to study the details of that writer's work.

5

Some Terms I Do Not Use and Why

The following terms often appear in discussions of inward and/or outward observation:

- ◆ Privileged access
- ◆ The external world
- ◆ Private and public
- ◆ Subjective and objective
- ◆ First-person and third-person

I do not use any of these terms, for the reasons explained in this chapter. To set the stage for these explanations, I first note various ways in which outward observation and inward observation are similar and various ways in which they are different.

Here are seven similarities:

1. *Attention.* In observing something either outwardly or inwardly, you focus your attention on it.
2. *Instances and types.* Both outwardly and inwardly, you can observe single instances and recognize types. Outwardly, for example, you can observe particular trees and also recognize species of trees or trees in general. Inwardly, you can observe particular headaches on specific occasions and also recognize headaches as a type.
3. *Description and analysis.* Both outward and inward observations can be described verbally and analyzed in various ways. Examples of analysis

that apply to both are counting instances of a recognized type and computing correlations between two recognized types.

4. *Potential for error.* Both outward and inward observations are subject to error, and for many of the same reasons. One reason is simple inattentiveness. Another is contamination by expectations, wishes, or preconceptions.

5. *Solo act.* Every observation, either outward or inward, is made by a single individual. In this respect, observing is like breathing and eating.

6. *Dependence on experience.* Both outward and inward observations involve experience. Behaviorist psychologists refuse to use the inward observation of experience in their research, confining themselves to outward observation of experimental subjects. But their own sensory experience is nevertheless involved in every outward observation that they make. All observation depends on the observer's experience, whether or not the observer's experience is its target.

7. *Cognitive contribution.* Both outward and inward observation can be informative. Both can support or undercut hypotheses. This point deserves emphasis because writers with behaviorist inclinations have long subjected inward observation to unjustified disparagement. There are many examples of synergy between outward and inward observation. Sitting in a chair with my legs crossed, I can look at the positions of my legs (outward observation) and simultaneously notice the pressure sensations caused by each leg pressing against the other (inward observation). A doctor's outward observations of a patient can be combined with the patient's inward observations of his unpleasant sensations to yield a medical diagnosis. Researchers look for correlations between what they can observe outwardly with brain-imaging or electrical recording techniques and what their experimental subjects observe inwardly.

Here are five differences:

1. *Role of spatial transmissions.* Outward observation depends on spatial transmissions from the target of observation to an observer's sense organs. For example, visual observation depends on the spatial transmission of light from the target of observation to an observer's eyes. Inward observation does not depend on spatial transmissions

from the target of observation to sense organs. The target of observation is a feature of the observer's experience to which she is able to direct her attention.

2. *Recognition of electronuclear objects.* Outward observation includes the ability to recognize different electronuclear objects—plants, animals, pieces of furniture, and so on. Inward observation does not include this ability. There is good reason to believe that inward observation is the observation of certain properties of a particular electronuclear object, namely the body of the observer, but this is not a simple recognition. It is a claim that many people have disputed.

3. *Role of the observer's experience.* In inward observation, the target of observation is a feature of the observer's experience. In outward observation, the target of observation is not experience, but sensory experience is an indispensable means of observation.

4. *Variety of relevant experience.* Only sensory experience that is associated with sense organs that are sensitive to spatial transmissions can serve as a means of outward observation. A much broader variety of experience can be observed inwardly.

5. *Number of observers who can observe a given instance.* In outward observation, a given object can be observed by many observers. Many people can see the same building, hear the same siren, feel the same piece of cloth, smell the same skunk, or taste the same peach. This is due to the role of spatial transmissions in outward observation. Objects radiate spatial transmissions in many directions. Many people in a variety of locations can intercept these transmissions with their sense organs and thus observe the object that is their source. In inward observation, a given instance of a feature of experience can be observed by only one observer, namely the person whose experience it is. Only I can observe my itches; only you can observe your aches.

I turn now to the problematic terms.

<>

Privileged access. It is sometimes said that a person has "privileged access" to his own experience. This is a way of describing the fact that a person has a type of access to his own experience that no one else has. I prefer the terms "inward access" and "inward observation," for two reasons. First, we normally use the word "privilege" to describe social arrangements, but the subject here is an innate characteristic of the human animal that does not depend on social arrangements. Second, in keeping with its allusion to social status, the phrase "privileged access" is often used to suggest access of an especially high quality. To describe this especially high quality of access, words like "infallible," "incorrigible," and "authoritative" are sometimes used. This claim of especially high quality is dubious. As I have already noted, inward observation is susceptible to many of the same errors as outward observation. Observation of both types can be done more or less attentively, more or less accurately, and more or less thoroughly.

<>

The external world. In *The Metaphysical Foundations of Modern Science,* E. A. Burtt gives a brief review of early-twentieth-century contributions to his subject in which he complains about "the continued uncritical use ... of traditional ideas like that of 'the external world'" (1954, 28). The uncritical use of this phrase continues undiminished to the present day.

The question is "External to what?" Occasionally, this phrase is used to mean everything external to a given person's body. That meaning is clear, but it is not the most common meaning, which includes everything that is visible. In his essay "Proof of an External World" G. E. Moore (1959, 145–146) uses his own hands as examples of objects in the external world. I once woke up shortly before the end of a colonoscopy procedure and found myself staring at a video screen that showed the glistening pink interior of my colon, lit up like a television studio. I was looking at a portion of the external world, as this phrase is most often used.

Someone who holds a body-plus view of human beings can understand "the external world" as the world external to the nonbodily component—everything in the world except Cartesian thinking things or Huxley's "mental steam," for example. Perhaps this is how the early users of this phrase understood it. But if all body-plus views are mistaken, this way of understanding "the external world" presupposes a false belief and should be avoided for that reason.

What about the answer "external to experience"? If experience consists of properties of a certain electronuclear system, then this answer involves an abnormal use of the word "external." It sounds odd to say "the world external to temperature" or "the world external to motion." If the idea is to refer to the world imaginatively stripped of some of its properties, "external" is not the right word.

Thomas Nagel, who uses the phrase "the external world" quite a bit, glosses it on one occasion as "the world external to our minds" (1995c, 98). This formula is unhelpful because the word "mind" is ambiguous and each way of understanding this word leads to one of the problems already mentioned. If "mind" is understood in a body-plus way, Nagel's formula makes the phrase "the external world" dependent on an implausible body-plus view. If "mind" is understood in a body-only way to mean a certain set or cluster of properties of the electronuclear body, Nagel's formula uses the word "external" in an abnormal manner.

In sum, the question "External to what?" does not have a good answer, and therefore the phrase "the external world" does not have a good use. Those who use it are either presupposing an implausible body-plus view of human beings, or misusing the word "external," or simply repeating a well-worn phrase without giving much thought to what they mean by it. If one could somehow force every writer who uses this phrase to provide a good answer to the question "External to what?" I suspect that its use would soon die out.

<>

Private and public. The phrase "publicly observable" is often used in connection with outward observation. The idea is that an outwardly observable object is publicly observable in the sense that it can be observed by many people. By contrast, inward observation is said to be "private." The trouble with this usage is that outward observation and inward observation are both public and both private, in different senses. The conventional contrast between "public" outward observation and "private" inward observation depends on an arbitrary focus on a public aspect of outward observation and a private aspect of inward observation. The complete comparison is more complex.

Inward observation is indeed private in the sense that only the person who has a given instance of a feature of experience can inwardly observe it. But inward observation is also public, in the sense that there are numerous features of experience that many people can inwardly observe by observing their own private instances of them. Three examples are feeling hungry, feeling sleepy, and having a visual representation of a bright red traffic light. In circumstances where there is reason to believe that the experiential states of two or more people are similar in some noteworthy respect, it is common for someone to verbalize that fact in a sociable manner. "Hot enough for you?" "That was scary!" "What a bore." It is true that not all features of experience are widely instantiated in many people. For example, few people have the opportunity to inwardly observe the weightless feeling of space travel or the inner voices of schizophrenia. But there are many features of experience that are widely instantiated in many people, and this fact gives the inward observation of experience a public aspect.

Outward observation is indeed public in the sense that many people can outwardly observe the same object. But outward observation is also private, in the sense that every case of outward observation is a solo act involving instances of experience in the observer that only the observer can inwardly observe. The experiential means of an outward observation is as private as any inward observation.

In sum, inward observation is private in a sense but also public in a sense, while outward observation is public in a sense but also private in a sense. It is misleading to say simply that one is private and the other is public.

<>

Subjective and objective. This is an extremely slippery pair of words. They are generally understood as contrasting with one another, but they are used to mark several different contrasts. Worse, they are often used in nebulous ways that make it difficult to be sure which of these contrasts the writer intends.

In one pair of senses, "subjective" and "objective" are used to mark the distinction between inward observation and outward observation. Inward observation is said to be subjective and outward observation is said to be objective. This usage is common in medicine. A symptom that is observed inwardly by a patient, such as pain or dizziness, is said to be subjective. A symptom that doctors can observe outwardly, such as a rash or the results of a blood test, is said to be objective.

In a second pair of senses, all experience is said to be "subjective," and "objective" means nonexperiential or existing apart from experience. What the word "subjective" *means* when it is used in this way is not clear to me; it seems to make the phrase "subjective experience" redundant, scarcely different from "experiential experience." In any case, there is a sense of "subjective" that is coextensive with "experiential."

In a third pair of senses, "objective" means "existing independently of the mind" or "mind-independent," and the contrasting sense of "subjective" is mind-dependent. This is similar to the second pair of senses, but not quite the same, because the word "mind" is more nebulous and potentially more inclusive than the word "experience." For example, memory content in its long-term storage format might be considered mind-dependent and hence subjective in this third sense, but it is not subjective in the second sense because it is not experiential.

It sometimes seems to me that a writer who says he is using "objective" in the sense of "existing independently of the mind" is actually using it to mean "existing independently of being observed or thought about" or "existing independently of anyone being aware of it." In this sense, everything that exists is objective and the contrasting sense of "subjective" applies to nothing—unless a view such as Berkeley's idealism is correct. Even thoughts exist independently of being thought about, because when you think you are thinking about something other than the thought you are having.

At the end of chapter 3, I raised a question about the relation between plenary electronuclear objects and their outward profiles. Suppose that a water molecule has properties that are neither outwardly observable nor experiential—a possibility to be discussed at length in coming chapters. Are these properties objective? The answer is yes if "objective" means independent of experience (second sense) or independent of the mind (third sense), but the answer is no if "objective" means outwardly observable (first sense). Because of this ambiguity, use of the word "objective" makes it extremely difficult to think and write clearly about the relation between plenary electronuclear objects and their outward profiles.

In chapter 4, I discussed the distinction between the entirety of animal experience and those features of experience that are inwardly observable. In the second and third senses of "subjective," all experience is subjective, but in the first sense, only inwardly observable experience is subjective. Because of this ambiguity, use of the word "subjective" makes it extremely difficult to think and write clearly about the distinction between the entirety of experience and the scope of inward observation.

In yet a fourth pair of senses, "objective" and "subjective" are used to characterize the grounds of a judgment. A subjective judgment is one that is influenced by personal factors such as tastes, prejudices, or financial self-interest. An objective judgment is one that is based entirely on impartial cognition and not influenced by any personal factors. Used in this way, "subjective" is often a term of disapproval while "objective" is commendatory. The notion that being "objective" is better than being "subjective" has some tendency to bleed over into the other senses of these words, which is inappropriate and potentially confusing. It creates a subliminal message that outward observation is somehow superior to inward observation, the nonexperiential is somehow superior to the experiential, and the mind-independent is somehow superior to the mind-dependent—all of which is baseless.

There are still other senses of "objective" and "subjective," but these four are the ones that bear on my subject.

The words "objective" and "subjective" play central roles in the work of two of the writers that I discuss in this book, Thomas Nagel and John Searle. I do not think that these words serve either of them well. For example, I suspect that Nagel's neglect of the distinction between inwardly observable experience and experience that is not inwardly observable, which I discussed in chapter 4, may be partly due to the ambiguity of "subjective." Other ways in which these terms seem to distort the thinking of these two writers will be noted later.

<>

First-person and third-person. Grammar books use these terms to classify sentences for the purpose of conjugating verbs. A sentence whose grammatical subject is "I"—the speaker—is a first-person sentence. A sentence whose grammatical subject is "you"—the person spoken to—is a second-person sentence. A sentence whose grammatical subject refers to anything other than the speaker or the person spoken to is a third-person sentence. The grammatical subjects of third-person sentences can take many forms, including pronouns like "he" and "she," proper names like "Abraham Lincoln" or "New York City," and descriptions like "the house across the street." One quirk of this terminology is that it gives the word "person" an unusual breadth. So-called third-person sentences can be about anything at all, not just people. The "you" of a second-person sentence can refer to a dog or a deity. It can even refer to a flower or a star, if the speaker is a poet. The "I" of a first-person sentence can refer to a robot, an extraterrestrial, or a farm animal, if the sentence appears in a work of fiction. Another quirk of this terminology is the arbitrariness of the numbering scheme. One could just as easily number these three grammatical categories in a different order. Both of these quirks could be avoided by using descriptions such as "speaker," "spoken to," and "other," but the conventional terminology is deeply entrenched and unlikely to be replaced any time soon.

Somewhere along the way, writers started using the adjectives "first-person" and "third-person" outside the context of grammar rules to qualify a wide variety of nouns, including all of the following: access, data, investigation, judgment, methodology, perspective, point of view, report, ontology. It is difficult to pinpoint what "first-person" and "third-person" are supposed to mean outside the grammatical context, because these terms have so little descriptive content. Talk of the first-person perspective or first-person point of view seems to mean something like inward observation or perhaps inward access in general. Talk of the third-person perspective or third-person point of view seems to mean something like outward observation.

One problem with this usage is that a person can observe himself both inwardly and outwardly. If I look at my hand, listen to my stomach growl, smell my armpits, feel the bumps on my skull, or taste my blood, I am observing myself outwardly. By the first-person/third-person terminology as it is normally used, I thereby have a "third-person perspective" on myself. That is an odd statement; the terms do not transfer well.

A second problem is that both "first-person perspective" and "third-person perspective" include so much. I'll focus on "third-person perspective." This term is typically taken to include not only the pure outward observation that is characteristic of the physical sciences, but also observation of human facial expressions and body language that involves the fellow-human factor as well as comprehension of verbal communications. It might even include all thinking about other people. David Papineau (2002, 139–140) uses the phrase "third-person phenomenal judgement" in a way that seems to include all beliefs about other people's experience, including mere suspicions and paranoid delusions. The mere existence of a broad term is not a problem, but the difficulty is that users of "first-person perspective" and "third-person perspective" tend to treat them, like their grammatical counterparts, as fundamental categories that do not need to be subdivided. This places an obstacle in the way of recognizing important distinctions within these broad categories, such as the partition of the "third-person perspective" into the pure outward observation that is characteristic of the physical sciences, observation of fellow humans that involves the fellow-human factor, comprehension of interpersonal communication, and arbitrary thoughts about the experience of other people. To keep track of these distinctions, we need a more refined vocabulary that is anchored in acquaintance with the subject being discussed, instead of terminological transplants from a grammar book.

A third problem is that the descriptive poverty of "first-person" and "third-person" leaves them open to a Wild West of disparate uses. Following are a few examples of the confusing proliferation of uses that has developed.

Video game manufacturers use the terms "first-person perspective" and "third-person perspective" to distinguish two screen views that a player can choose between. In the first-person perspective, the player sees a landscape from the perspective of the game character that he controls; all he sees of his game character is a pair of hands holding a weapon. In the third-person perspective, the player sees a landscape from a perspective that is not the perspective of any character in the game; he sees the entire body of his game character. Notice that both of these views involve only outward observation. Where academic writers typically use these terms to distinguish outward observation from inward observation, video game manufacturers use them to distinguish two vantage points from which outward observation can be conducted. When you play a video game, inward observation is not on the menu.

John Searle uses the adjectives "first-person" and "third-person" to qualify the noun "ontology." The resulting phrases "first-person ontology" and "third-person ontology" can only be understood within the context of Searle's view, which I discuss in chapter 13.

Herbert Feigl (1967, 70) uses the phrase "the first-person data of direct experience" to refer to inwardly observable features of experience. However, Gualtieri Piccinini (2009) uses the phrase "first-person data" in a way that pointedly excludes inwardly observable features of experience, applying instead to the responses that participants in psychology experiments make to questions about their experience. Such responses can take the form of verbal reports, acts of pointing at pictures, or acts of pressing a button. Each of these uses of "first-person data" seems reasonable in itself, but having them both in use is confusing.

Daniel Dennett would presumably object to Piccinini's use of "first-person data" for the same reason that he objects to the use of "first-person investigations" in the following passage:

> And when Hurlbut and Heavey say 'For example, first-person investigations often rely on questions such as "What were you thinking when you ...?" or "How were you feeling when you ...?" ' it apparently does not occur to them that these aren't first-person investigations; they are third-person investigations of the special kind that exploit the subject's capacity for verbalization. (Dennett 2003, 23–24)

51

This disagreement is due largely to the confusing nature of the terminology. Psychology experiments in which the participants are asked to attend to their thoughts and feelings and then describe them to a professional researcher can reasonably be called first-person investigations, because they involve inward observation. They can also reasonably be called third-person investigations, because they involve interpersonal communication. Arguing over whether such an experiment is a first-person investigation or a third-person investigation is like arguing over whether a brand of beer tastes great or is less filling. In the grammatical context, a sentence cannot be both first-person and third-person. But when these adjectives are applied to psychology experiments, they need not exclude each other. A psychology experiment is a complex social undertaking that can be first-person in one sense and third-person in another sense.

Dennett is a heavy user of "first-person" and "third-person" whose work contains a variety of interesting examples. His uses of these terms tie in with his attempts to discredit inward observation.

In the following passages, Dennett coins the phrase "first-person-plural perspective":

> In fact, just about every author who has written about consciousness has made what we might call the *first-person-plural presumption*: Whatever mysteries consciousness may hold, *we* (you, gentle reader, and I) may speak comfortably together about our mutual acquaintances, the things we both find in our streams of consciousness. (Dennett 1991, 67; his emphasis)

> The standard perspective adopted by phenomenologists is Descartes's first-person perspective, in which I describe in a monologue (which I let you overhear) what I find in my conscious experience, counting on us to agree. I have tried to show, however, that the cozy complicity of the resulting first-person-plural perspective is a treacherous incubator of errors. (Dennett 1991, 70)

These passages combine a good point with an erroneous suggestion. The good point is that many writers do dubiously presume that their readers are just like them in various respects. "Speak for yourself," I often find myself thinking. The erroneous suggestion is that attending to and reporting features of your own experience is somehow tantamount to making such presumptuous claims about humanity in general. Suppose I keep a record of certain abnormal sensations, hoping to help my doctor understand my medical condition. Or suppose I keep a diary of my dreams and look for themes and patterns in it. These are examples of inward observation that involve no generalization of my findings to other people. Insofar as Dennett equates "first-person perspective" with "first-person-plural perspective," he is blurring an important distinction.

Another interesting example is illustrated by the following pair of passages:

> The Behaviorists were meticulous about avoiding speculation about what was going on in *my* mind or *your* mind or *his* or *her* or *its* mind. In effect, they championed the *third-person perspective*, in which only facts garnered "from the outside" count as data. (Dennett 1991, 70; his emphasis)

> Good scientist that Descartes was, he appreciated the value of intersubjectivity, and the ways that science has of canceling out the idiosyncrasies of individual investigators so that all can participate together in a shared inquiry, the "third-person" approach of scientific method. (Dennett 2005, 28)

The first of these passages associates "third-person" with outward observation, but the second associates "third-person" with shared inquiry. This is confusing because outward observation and shared inquiry are two different and mutually independent things. On the one hand, each of us uses solo, unshared outward observation all day long to make our rounds in the world. Is this the "third-person perspective," because it is outward observation, or not, because it involves no sharing? On the other hand, there can be shared inquiry that has nothing to do with outward observation. A prime example is the community of professional mathematicians, who share and compare their mathematical thoughts. Is this the "third-person perspective," because it involves sharing, or not, because the shared material is not obtained by outward observation? The results of outward observation are shareable, as research in the physical sciences illustrates, but so are many other things. People can also share hypotheses, computations, arguments, worries, and emotional reactions, all of which are accessed inwardly. Is Dennett suggesting that "third-person" outward observations can be shared and compared while "first-person" inward observations cannot? If so, he is mistaken.

Here is one final passage from Dennett:

> Ignoring all tempting shortcuts, then, here is the *neutral* path leading from objective physical science and its insistence on the third-person point of view, to a method of phenomenological description that can (in principle) do justice to the most private and ineffable subjective experiences, while never abandoning the methodological scruples of science. (Dennett 1991, 72; his emphasis)

If "the third-person point of view" signifies the use of outward observation, then this sentence seems to say that physical scientists insist on using only outward observation as a matter of methodological scruple. This is importantly not the case. For the objects that physical scientists study, outward observation is the only game in town. One can't insist on a scruple unless there is an unscrupulous alternative that one can voluntarily refrain from, and here there is none. If you are going to study molecules and meteors, then you are going to study them by means of spatial transmissions that travel from those objects to your sense organs. For such objects—which means for most objects in the world—outward observation is a methodological prison, not a methodological scruple. Dennett himself has a behavioristic or quasi-behavioristic "scruple" regarding the study of consciousness. In this sentence, he seems to be trying to win respect for this "scruple" by claiming that it is endorsed by the community of physical scientists. If that is what he is doing, he is making a mistake. Physical scientists do have methodological scruples, but limiting themselves to the means of observation that they are already limited to by natural circumstance is not one of them.

Using the physical sciences as a methodological model for the study of consciousness is a tricky business. It is true that research in the physical sciences does not use inward observation. But it is also true that research in the physical sciences exploits every available means of observation for the objects under study. It is not possible to study consciousness in a way that emulates the physical sciences in both of these respects. One of the means of observation that is available for the study of consciousness is inward observation. Therefore, if you use every available means, you use inward observation, and if you don't use inward observation, you don't use every available means. Dennett thinks of his behavioristic or quasi-behavioristic "scruple" as making the study of consciousness like the physical sciences, in not using inward observation. He does not note that it also makes the study of consciousness unlike the physical sciences, in refusing to use an available means of observation. Alternatively, one can study consciousness in a way that is like the physical sciences in using every available means of observation, and therefore unlike the physical sciences in that it uses inward observation. Neither of these ways of studying consciousness is "more like" the physical sciences than the other. One can debate which of them is more reasonable, but to me the answer is clear. If our goal is to understand the things we study as fully as we can, then we should exploit every available means of observation. It makes no sense to renounce a means of observation in an arena where it is available, just because there is another arena in which it is not available.

I do not use any of the terms I have criticized here—privileged access, external world, private and public, subjective and objective, first-person and third-person—because the ambiguities and misconceptions that they harbor threaten the standard of clarity that I would like to maintain.

6

Brains

Many writers claim that all the features of an animal's experience are properties of the animal's brain. Some claim that particular features of experience are properties of particular parts or regions of the brain. These claims are problematic for reasons that I explain in this chapter. The quadrants view is compatible with ascribing features of experience to brains, but I think it makes more sense to ascribe all the features of an animal's experience to the whole living electronuclear system.

The trouble with ascribing features of experience to the brain is that their existence also depends on various other parts of the body. In a frequently discussed thought experiment, super-knowledgeable scientists of the future maintain a "brain in a vat," which they supply with sophisticated electrical stimulation that gives it a wealth of life-like experience. The absurdity of this thought experiment tends to go unnoticed, perhaps because "brain in a vat" is such a catchy phrase. A real human brain can generate experience only if it is continuously supplied with oxygen and nutrients while metabolic waste products are continuously removed from it. This means that it has to be served by a circulatory system, which in turn has to be connected to modules that regulate the contents of the circulating fluid. Just possibly, all this could be achieved with man-made laboratory equipment instead of a living human body. But one would need an artificial heart, lung, liver, kidney, and much more. Moreover, it seems that the circulating fluid would have to be genuine human blood—no substitute is possible here—and that you would have to retain the intricate natural network of tiny blood vessels that brings blood to and from every cell of the brain. A stand-alone human brain in a vat of brain-friendly fluid could not generate experience, no matter what kind of electrical stimulation scientists might provide. It could not function at all, except as a museum exhibit.

Hilary Putnam describes the "brain in a vat" thought experiment in the following way:

> Here is a science fiction possibility discussed by philosophers: imagine that a human being (you can imagine this to be yourself) has been subjected to an operation by an evil scientist. The person's brain (your brain) has been removed from the body and placed in *a vat of nutrients* which keeps the brain alive. (Putnam 1981, 5–6; my emphasis)

Here Putnam seems to imagine a living, functioning human brain getting the nutrition it needs from ambient fluid rather than from a finely branching circulatory system driven by a pump. Putnam claims that this arrangement "violates no physical law" (1981, 7) and that it is "a state of affairs which is compatible with the laws of physics" (1981, 15). These claims are doubtful in the extreme. A brain not served by a circulatory system is like a mill wheel in a stagnant pond. At best, the set-up Putnam describes would allow the brain's outermost cells to absorb nutrients while gradually polluting the ambient fluid with their waste products. Most of the brain's cells would quickly die and all of them would be dead very soon.

Thomas Nagel writes, "I could lose everything but my functioning brain and still be me" (1986, 40). Maybe so, but this bold declaration cries out for a footnote: the adjective "functioning" implicitly brings in a swarm of nouns. If someone were to suddenly lose all of his body except his brain, he would have to suddenly "find" an enormously complex artificial support system to keep his thoughts and feelings going. There can be no such thing as a normally functioning stand-alone brain. Accordingly, any statement that "the brain" thinks, feels or sees has to be understood in a highly contextual way or else rejected as a falsehood. It is the whole living animal that does these things, with the brain playing a leading role.

<>

There is a wealth of evidence that particular features of experience have special relationships to particular regions of the brain. Visual experience makes a good example. Brain imaging studies of human beings and electrical recordings from the neurons of nonhuman animals show that the patterns of activity in certain regions of the brain, but not in other regions, change quickly in response to changes in the light that is striking the animal's eyes. In *The Quest for Consciousness*, Christof Koch describes what he calls the "network wave" of light-triggered neural activity, which is observed to rush along the optic nerves to the lateral geniculate nucleus in the thalamus, from there to the primary visual cortex in the occipital lobe, and then to various parts of the temporal and prefrontal cortex. From these observations one can infer that these parts of the brain play special roles in the generation of visual experience.

Koch describes the use of binocular rivalry experiments with monkeys to learn more about what these roles are. It is known from the human case that when a barrier is placed between a person's two eyes and each eye is presented with a different picture, the person's visual experience switches back and forth between the two pictures, rather than blending them together. Presumably, the extreme difference between the two pictures thwarts the brain's normal operation of smoothly integrating inputs from the two eyes, thus forcing it to favor the input from one eye or the other. In the experiments that Koch describes, a monkey is trained to indicate by pushing levers whether it sees a picture of an orangutan face or a geometrical design known as a sunburst pattern. The monkey is then presented with both pictures simultaneously, one to each eye, and the electrical activity of various neurons that are in the path of the light-triggered network wave is monitored. What these experiments show, in brief, is that when the monkey pushes a lever to indicate that its visual experience has just switched from one picture to the other, there is not much change in recorded activity in the regions that the network wave travels through first, including the optic nerves, the lateral geniculate nucleus in the thalamus and the primary visual cortex in the occipital lobe, but there are major changes further along the path, in the temporal and prefrontal lobes. This result suggests that all the nervous tissue from the retinas to the primary visual cortex does preparatory processing, after which the active regions of the temporal and prefrontal lobes generate and house the content of the monkey's current visual experience, including such features as color, shape, motion, and object recognition.

There is a temptation to conclude from this sort of experiment that the features of a monkey's visual experience are properties of these regions in the temporal and prefrontal lobes. This is a mistake. It is a mistake because the experiment shows only that these regions contribute what distinguishes these features of experience from other features of experience. It tells us nothing about the source of what these features of experience have in common with other features of experience, namely their experiential-ness. And there are good reasons to believe that their experiential-ness has dependencies on other regions of the brain as well as on the body at large, as already noted. You can calm down all these vision-specific regions of the brain by blindfolding a person or putting him in a very dark room, but such a light-deprived individual remains in an awake, experiential state, perhaps marked by idle thoughts or itchy feet.

Koch makes this point by distinguishing "specific factors" from "enabling factors" (2004, 88) or "background conditions" (2004, 114). He describes this distinction in the following passage:

> Enabling factors are tonic conditions and systems that are needed for any form of consciousness to occur at all, while specific factors are required for any particular conscious percept, such as seeing the glorious, star-studded alpine night sky....
>
> Some authorities argue for the need to distinguish between the content of consciousness, on the one hand, and the "quality of being conscious" or "consciousness as such" on the other. This distinction maps straightforwardly onto my classification. (Koch 2004, 88)

Some of the "enabling factors" that Koch mentions are glia cells, brain stem activity of a certain sort, and activity of a certain sort in the intra-laminar nuclei of the thalamus. To this I add the following two remarks. First, there are also "enabling factors" in parts of the body outside the brain, especially the blood and the various organs that keep the blood fresh and flowing. Second, all these so-called "enabling factors" or "background conditions" are every bit as necessary to the resulting experience as the "specific factors" are. It would be more evenhanded to contrast "specific factors" with "generic factors." You can't have colorful visual experience without a source of the specific color features, but equally you can't have colorful visual experience without a source of a generic experiential state.

There is a pertinent analogy between features of experience and beverages. There are many different beverages—coffee, tea, milk, fruit juices, soft drinks, wine, beer, and so on. These beverages have different flavors, different textures, and different temperatures; they are very different from one another. Yet they have something important in common: they all consist mostly of water. Parallel remarks apply to features of experience. There are many features of experience. They are very different from each other. Yet they have something important in common: they are all experiential. In discovering outwardly observable nerve activity that happens if and only if a person has a particular feature of experience, one is locating cells that help to contribute what is distinctive about that feature of experience. But there is more to a feature of experience than what is distinctive about it. There is also its experiential-ness. Just as every beverage consists mostly of water, so every feature of experience has dependencies that it shares with every other feature of experience. I can understand how someone who is mainly interested in beverage flavors might consider water an "enabling factor" or a "background condition," but if the subject is whole beverages and not just beverage flavors, then it is not right to relegate the water in this way. Likewise, if the subject is experience as a whole and not just what makes one feature of experience different from others, it is not right to relegate the generic quality of being experiential. Parts of the brain and parts of the body beyond the brain that contribute to the generic quality of being experiential are as much part of the locus of a given feature of experience as are the parts of the brain that contribute the specific qualities that distinguish that feature of experience from others.

<>

In the film *Monty Python and the Holy Grail*, King Arthur lops off the Black Knight's arms and legs, yet the Black Knight continues to taunt him. This hilarious scene illustrates the serious point that a person without arms and legs can lead a richly experiential life. Indeed, there is even more that the Black Knight could lose without ceasing to have experience. A modern-day King Arthur with surgical training could remove a lung and a kidney, perform an appendectomy and a tonsillectomy, and whittle away a bit more miscellaneous tissue before setting the Black Knight up in an assisted living facility where he could continue to speak ill of the king for years to come.

One might cite this as a reason why it is wrong to ascribe features of experience to a whole living body. Much of the body is unnecessary, even irrelevant to having experience, one might claim. There is something to this, of course, but the problem is that there is no good way to define a system smaller than the whole body that has all and only the indispensable parts. A stand-alone brain is much too small, and once you include the indispensable blood, heart, liver, lung, and kidney, there is no clean boundary to be drawn except at the surface of the skin. This is why I think it is best to say that features of experience are properties of the whole living electronuclear system.

Because so many of the writers I discuss focus exclusively on the brain and ignore the contributions of other parts of the body, I find it convenient in the chapters ahead to "go with the flow" and say that features of experience are properties of the brain. These statements should always be understood in the highly contextual sense that features of experience are properties of a system that includes a brain and a host of indispensable supporting modules. Some of these supporting modules can be prosthetic, but any such brain-supporting system must have major similarities to a natural animal.

7

The Quadrants

A human being—a human electronuclear system—has many outwardly observable properties and many experiential properties. Do these two sets of properties overlap? In other words, are there experiential properties that are outwardly observable? I find it plausible that there is no overlap, for two reasons. First, no matter how much advanced technology is used, outward observation of a living body does not seem to reveal its experiential properties. Second, it is hard to imagine how outward observation of a living body could reveal its experiential properties; inward observation seems uniquely suited to that purpose. Some writers take a different view. For example, according to some so-called identity theories, *every* experiential property is outwardly observable. Chapters 20 and 21 examine identity theories and give reasons to reject them. In this chapter I proceed on the basis of the plausible belief that no experiential property is outwardly observable.

At the end of chapter 3, I introduced the distinction between plenary electronuclear objects and their outward profiles. I then asked how the two are related: are plenary electronuclear objects propertywise plumper than their outward profiles, or not? I can now begin to answer this question. If human electronuclear systems have experiential properties that are not outwardly observable, then at least these electronuclear objects are propertywise plumper than their outward profiles. The same holds for all nonhuman animals that have experience. Any electronuclear system that has experiential properties is propertywise plumper than its outward profile. My answer for electronuclear objects in general comes later in this chapter.

The union of the class of inwardly observable properties and the class of outwardly observable properties is the class of properties that are observable either inwardly or outwardly. To simplify the terminology a bit, I will call this union the class of observable properties. The class of *observable properties* and the class of *properties that are not observable* together constitute a logically exhaustive dichotomy of all properties of all electronuclear objects. The class of *experiential properties* and the class of *nonexperiential properties* together constitute another logically exhaustive dichotomy of all properties of all electronuclear objects. Crossing these two dichotomies yields four classes of properties, which can be arranged in a 2 x 2 grid as shown below.

	Observable	Not Observable
Experiential	(Inwardly) observable experiential properties "the observable experiential quadrant"	Experiential properties that are not observable "the unobservable experiential quadrant"
Not Experiential	(Outwardly) observable nonexperiential properties "the outward quadrant"	Nonexperiential properties that are not observable "the remote quadrant"

The four classes of properties in this grid are what I call the quadrants of the corporeal. In each cell of the grid, the phrase in quotation marks is a convenient way to refer to the quadrant represented by that cell. I have shaded the cell that represents the outward quadrant—the class of all outwardly observable properties of all electronuclear objects. The class of all properties of all electronuclear objects is represented by the four cells together—the whole grid. Thus, there are three quadrants that could conceivably play a role in making an object propertywise plumper than its outward profile—the observable experiential quadrant, the unobservable experiential quadrant, and the remote quadrant.

Let's take a more detailed tour of this grid. The tour starts at the outward quadrant, continues with a combined discussion of the two experiential quadrants, and ends at the remote quadrant.

The outward quadrant contains all real properties that human beings or other observers, located anywhere in the vast expanse, can find evidence for by means of sense organs that are sensitive to spatial transmissions. It contains properties that we discover through the everyday use of our sense organs and also properties that we discover through the technologically assisted uses of our sense organs that are characteristic of present-day research in the physical sciences. The same holds for technologically advanced observers elsewhere in the vast expanse, if there are any.

This quadrant also contains properties whose actual discovery must await future developments in the means of outward observation. These could be either future technological developments within a society of observers or future developments in the evolution of observers.

If a cosmic plague were to kill every observer in the vast expanse, the set of outwardly observable properties would not change. A property is in this quadrant if and only if some process can place evidence of it into some spatial transmission or some set of spatial transmissions. If there is evidence of the property in spatial transmissions, then a sufficiently sophisticated observer can intercept the transmissions, extract the evidence, and have an outward cognitive encounter with the property. The competence of currently existing observers to receive, process, and interpret the spatial transmissions is not relevant to the classification.

The two experiential quadrants together include all experiential properties of all electronuclear systems. Electronuclear systems that lead experiential lives could reside on many planets throughout the vast expanse. However, all the experiential lives that we know of run their course right here on planet earth, so I will focus on our home planet.

Planet earth has existed for between four and five billion years. Living organisms have been evolving on it over most of this time. Animals having sense organs and a capacity for movement have been evolving at least since the Cambrian explosion of marine invertebrates over a half billion years ago. Among these sensing and moving animals are human beings. We human beings have an extensive repertoire of experiential properties that are tightly integrated with our sense organs, our neural anatomy, and our capacity for movement. If one accepts this set of facts, then it follows with near certainty that repertoires of experiential properties have been evolving as an integral part of evolving animal systems for a very long time. Therefore, experiential properties are widespread across the animal kingdom.

While the widespread existence of animal experience is a near certainty, the detailed nature of the experience of other animals is impossible to pin down. We cannot observe it outwardly. We cannot have it, and therefore we cannot observe it inwardly. This limits us to educated guesswork based on the inward observation of our own experience and the outward observation of the anatomy, physiology, and behavior of these other species. If we have retinal cells tuned to three frequency bands of light, while dogs have retinal cells tuned to only two frequency bands, and birds have retinal cells tuned to four frequency bands, one can surmise that our repertoire of color sensations is greater than that of dogs but less than that of birds. Donald R. Griffin (1981 and 1992) has sought to characterize animal thought processes based on observations of communication and problem-solving behavior. Jonathan Balcombe (2006) has sought to understand the extent to which animals have fun. For example, he sees a parallel between people sliding down playground slides and ski trails, penguins and bears sliding down snowy inclines, otters sliding down muddy river banks, and ravens and crows sliding down slanting rooftops and the rear windows of automobiles. All this engagement with slippery slopes, he argues, serves no purpose other than having a good time. Temple Grandin (2005) believes that autistic people such as her are experientially more similar to dogs, cats, and farm animals than normal people are. She thinks that nonhuman animals and autistic humans are aware of more perceptual detail because they lack the conceptual overlay that is supplied by a normal human frontal cortex. I find such speculations intriguing and persuasive, but they do not give us a detailed understanding of the experience of other animals. The near certainty that there is a lot of experience spread across the animal kingdom is combined with an insurmountable obstacle to appreciating its detailed nature.

In chapter 4 I noted that the human capacity for inward observation of experience has little or no counterpart in other animal species, and that it is limited in certain ways even in humans. Some features of experience are open to inward observation while others are not. This is the basis of the distinction between the two experiential quadrants. The observable experiential quadrant contains all features of experience that are inwardly observable. The unobservable experiential quadrant contains all features of experience that are not inwardly observable. Features of experience that are inwardly observable in a sense or to a degree are in a "gray area" at the boundary between the two experiential quadrants.

The distinction between the observable experiential quadrant and the unobservable experiential quadrant lines up roughly, but only roughly, with the distinction between human experience and nonhuman experience. There are several reasons why the two distinctions do not coincide. First, as explained in chapter 4, there are features of human experience that elude the human capacity for inward observation. Second, there is some capacity for inward observation in some nonhuman species. Even if no other contemporary species has any capacity for inward observation, some of our extinct pre-human ancestors must have had a capacity for inward observation that was not too different from ours. Third and finally, there is the possibility of extraterrestrial beings that match or even surpass us in the ability to observe experience inwardly. The distinction between the two experiential quadrants is based on accessibility by inward observation, not on species membership.

The remote quadrant contains all properties that are neither experiential nor outwardly observable. I call this quadrant the remote quadrant because any properties in it are extremely remote in a cognitive sense. The properties in every quadrant are cognitively remote to one degree or another, but those in the remote quadrant are the most remote, in the sense explained below.

Inwardly observable experiential properties are not cognitively remote at all for the creature that can inwardly observe them, but they are cognitively remote for other creatures that cannot inwardly observe them.

Experiential properties that are not inwardly observable are cognitively remote for all creatures, including the creatures that have them. However, for human beings, who can inwardly observe many of their own experiential properties, the cognitive remoteness of the experiential properties of other animals is tempered by the fact that they are experiential properties. This general resemblance to the experiential properties that we can observe in ourselves gives us a cognitive grip on them.

All outwardly observable properties are cognitively remote in the sense that we can only encounter them indirectly, by way of spatial transmissions and the sensory experience that spatial transmissions trigger in us when they strike our sense organs. However, the cognitive remoteness of outwardly observable properties is tempered by the fact that they have evidentiary effects on the spatial transmissions and, through the spatial transmissions, on an observer's sensory experience. Their evidence-delivering causal connection to our sensory experience gives us a cognitive grip on them.

Remote properties are related to human experience in none of these ways. They do not have the kind of resemblance to human experience that nonhuman experiential properties have. And they do not have the evidence-delivering causal connection to spatial transmissions that outwardly observable properties have. They are either incapable of having any causal connection to spatial transmissions or else their only possible causal connections to spatial transmissions do not produce evidentiary effects. All we have to characterize remote properties is the following set of very general and negatively characterized similarities to properties in the other quadrants:

+ Remote properties are like experiential properties in *not being outwardly observable.*

- Remote properties are like unobservable experiential properties in *not being observable at all, either outwardly or inwardly.*
- Remote properties are like outwardly observable properties in *not being experiential.*

This is something, but it is not much. Notice, in particular, that all three of these points apply uniformly to all remote properties. There is nothing here to distinguish one remote property from another.

<>

The crucial question concerning remote properties is whether there are any. The grid of quadrants shows that the existence of remote properties is a logical possibility, but it gives no reason to think that any electronuclear object actually has remote properties. It is also a logical possibility that all properties of all electronuclear objects are either outwardly observable properties or experiential properties. The remote quadrant would then be empty. Which of these two logical possibilities is the case? Are there real properties of real electronuclear objects that are neither outwardly observable nor experiential, or is the remote quadrant empty? The rest of this chapter is devoted to this question.

According to the view known as panexperientialism, extremely tiny objects such as electrons and molecular nuclei have experiential properties, albeit of a type that we are not familiar with. From this it follows that these experiential properties pervade the electronuclear fabric of all electronuclear objects. In chapter 12 I explain why I find panexperientialism implausible. Here I will assume, provisionally, that panexperientialism is wrong, and that many electronuclear objects—rocks, chairs, water molecules, electrons—do not have any experiential properties. All the properties of such objects are thus in the bottom row of the 2 x 2 grid, which I reproduce here:

	Observable	Not Observable
Not Experiential	(Outwardly) observable nonexperiential properties	Nonexperiential properties that are not observable
	"the outward quadrant"	"the remote quadrant"

Recall that the classification of a given property as outwardly observable does not depend on the existence of observers, but only on the existence of an evidentiary causal connection to one or more of the spatial transmissions on which outward observation depends. A nonexperiential property of an electronuclear object is a remote property if it does not have that kind of causal connection to any spatial transmission. Thus, the division of nonexperiential properties into the outwardly observable and the remote depends entirely on their causal relations to the existing spatial transmissions; it is not determined by the nature of the properties alone.

For an illustration of this point, consider the following speculative scenario. Some cosmologists have speculated that our vast expanse is one of many that are somehow stuck together like a frothy mass of bubbles. Suppose that this is the case. Suppose in addition that there exists a vast expanse that contains no spatial transmissions, but does contain some objects that do not have any experiential properties. These objects would not be electronuclear objects, because it is in the nature of electronuclear objects to send out the various spatial transmissions that human scientists use to study them. Let's call these strange objects x-blocks. Being in a vast expanse, x-blocks would have some of the same properties that electronuclear objects have in our vast expanse. Size and shape are two obvious examples. The size and shape of electronuclear objects in our vast expanse are outwardly observable properties. However, the size and shape of x-blocks in their vast expanse are remote properties. *All* properties of x-blocks are remote properties, due to the absence of spatial transmissions in their vast expanse. This example shows how the classification of a given property as outwardly observable or remote can shift if there is a change in the available supply of spatial transmissions.

Now let's set x-blocks aside and focus on our vast expanse with its electronuclear objects and spatial transmissions. All electronuclear objects that are devoid of experiential properties have many outwardly observable properties. Therefore, one of the following must be true:

- *All* the properties of all of these wholly nonexperiential electronuclear objects are outwardly observable. None of these objects has any remote properties.
- *Not all* the properties of all of these wholly nonexperiential electronuclear objects are outwardly observable. At least one such object has at least one remote property.

Following are four considerations that together make a strong case for the "Not all" alternative.

First, I know of no argument for the "All" statement. The history of the physical sciences shows that we can learn a great deal about electronuclear objects through outward observation, but that fact does not address the question at hand. No matter how much we learn through outward observation, the possibility remains that the outward quadrant and the remote quadrant are both well stocked. Like a rich philanthropist, nature might make many properties available to outward-observation-based scientific investigations while keeping other properties in secret accounts. Many people tacitly assume the truth of the "All" statement. This is the habit of outwardism, which I discuss in chapter 8. But it is just a habit, not a well-founded belief.

Second, think of typical electronuclear systems that do have experiential properties—human beings, dogs, bats. The experiential properties of human beings and other animals are not outwardly observable. Here, then, we have examples of electronuclear systems whose plenary nature is partially inaccessible by outward observation. *How is this possible?* There must be a barrier of some kind that keeps outward observation from making cognitive contact with the experiential properties. Whatever this barrier is, the portion of plenary nature that is "behind" it might also include nonexperiential properties.

In the following passage, Owen Flanagan seems to say that the inaccessibility of experiential properties by outward observation is explained by the fact that we have a different kind of access to (some of) them:

> We grasp facts about the brain through our sensory organs, typically with the help of sensory prostheses: MRI, CAT, and PET scanners. On the other hand, we are acquainted with consciousness by way of direct, internal, reflexive biological hookups to our own nervous systems.... The nature of our access to what we are made of and to how we function as complex biological systems is different in kind and provides different information than does our first-person, on-line hookup to ourselves. The biological fact that each of us possesses a direct, first-person, reflexive hookup to one and only one body explains how the most complete explanation from the physical point of view does not capture what it is like to be me. Only I can capture that. Only I am hooked up to myself in the right sort of way....

69

> The important point is that we possess a good naturalistic explanation of our inability to capture the phenomenology of what it is like to be each one of us from the objective point of view. The hookups, the epistemic access relations, are essentially different in the first-person and in the third-person. (Flanagan 1992, 117–118. There is an earlier version of this passage in Flanagan 1991, 343.)

This is not right. Two doors, no matter how different they are from each other, need not lead into different rooms. Likewise, two different epistemic access relations need not differ completely or indeed at all in the properties that they give us access to. For example, one can discover the shapes and sizes of many electronuclear objects by sight or by touch. One can even do so with sonar if one has the right equipment. In this case there are properties that are accessible through three different epistemic access relations. The fact that human beings are capable of both outward observation and inward observation merely sets up the question of how the respective scopes of these two types of observation are related to each other. It does not determine the answer to this question, or even bear upon the answer. If it is a fact that there is no overlap, that outward observation provides no access to any of the experiential properties of a human electronuclear system that a person can inwardly observe, then there must be a barrier of some kind that limits the scope of outward observation of human electronuclear systems. And a barrier that shields experiential properties from outward observation might also shield some nonexperiential properties from outward observation.

Note that this second consideration depends on experiential properties not being outwardly observable. An identity theorist who maintains that all experiential properties are outwardly observable will not accept this premise. The critique of identity theories presented in chapters 20 and 21 is therefore an important supplement to this second consideration.

Third, if there is a barrier of some kind that keeps outward observation from making cognitive contact with some portion of the plenary nature of electronuclear objects, then one must ask what kind of barrier this is and why it exists. Here is a plausible explanation for the existence and nature of such a barrier. The spatial transmissions that are our means of outwardly observing electronuclear objects have a nature that is distinct from and independent of the nature of the outwardly observed electronuclear objects. Consequently, the various properties of electronuclear objects might differ in their capacity to have effects on the spatial transmissions. Some properties might be capable of having evidentiary effects on the spatial transmissions, others might not be capable of having effects that carry evidence of the properties that cause them, and still others might not be capable of having any effects at all on the spatial transmissions. Outward observers, being wholly dependent on the spatial transmissions, are able to outwardly observe only those properties of the observed electronuclear objects that can have evidentiary effects on the spatial transmissions.

It is illuminating to reflect on this situation from an evolutionary perspective. The atmosphere and the oceans of the early earth were traversed by spatial transmissions of all the types that outward observers depend on today. There were electromagnetic waves, compression waves, diffusing molecules, fields of electric force. The first squishy micro-blobs of life had no sense organs; they were mere smidgens of electronuclear fabric with an ability to persist and reproduce. Being bombarded by the spatial transmissions, these little living things were able to begin evolving systems for exploiting the spatial transmissions in a way that improved their reproductive success. There was a lot of potential for improvement, because the spatial transmissions contain evidence about the environment that a living thing can use to find what it needs and avoid what can harm it. Over eons, outward-oriented sense organs such as eyes, ears, and noses developed. Several billion years of evolution have perhaps brought things to the point where we human beings can extract from the spatial transmissions all the evidence that they contain concerning properties of the transmitting objects. But we will never be able to extract from the spatial transmissions what is not in them. And there is no reason to think that what is in the spatial transmissions does justice to the plenary nature of electronuclear objects, because the spatial transmissions and the electronuclear objects are different things with independent and disparate natures. The electronuclear objects exist. The spatial transmissions exist. But there is no "pre-established harmony" between them. They were not designed to work together in the service of human cognition. Our outward study of electronuclear objects thus depends entirely on our exploitation of something other than those objects, namely the spatial transmissions. The extent to which these spatial transmissions can give us access to the nature of the electronuclear objects from which they come is a matter of happenstance.

Here is a simple analogy. Imagine that there is a zoo on an island. You have the task of transporting animals from the zoo to the mainland. The only tool at your disposal is a canoe. The canoe can carry cats and dogs, and maybe sheep and goats if you're careful, but it cannot carry cows and horses, let alone elephants and giraffes. You are in this predicament of partial capability because canoes were not designed to be general-purpose animal carriers. The canoe just happens to be the available tool. It carries what it carries. You do with it what you can. Likewise, the spatial transmissions were not designed to be general-purpose evidence carriers concerning the nature of electronuclear objects. They just happen to be the media that are available for studying electronuclear objects. They convey what they convey. We learn what we can by means of them. Sense organs and the capacity for outward observation evolved to take advantage of the spatial transmissions that exist. The spatial transmissions that exist are one thing, the electronuclear objects that exist are something else. Their relationship is incidental, like that of the canoe and the animals on the island. Given this set of facts, it is improbable that the spatial transmissions can reveal to curious observers the plenary nature of electronuclear objects.

Imagine that you are an acorn, or a water molecule. You have no thoughts to communicate, of course, but you do have a plenary nature. There are scientists out there eager to know all about you. But your only communication link to them is the spatial transmissions. The spatial transmissions have their own nature, as ancient and as adamant as your own. Do they provide an adequate means for conveying your plenary nature to curious researchers? It would be an improbable accident if they did. Those inquiring humans must make do with what the spatial transmissions can convey. From their point of view, some of your properties are outwardly observable and some are remote.

Defense Secretary Donald Rumsfeld memorably declared, "You go to war with the army you have." Likewise, outward observers of electronuclear objects do their observational work using the spatial transmissions that they have. There is no reason to think that these spatial transmissions are consummate cognitive tools, and good reason to suspect that they are not. The spatial transmissions are to the plenary nature of electronuclear objects as the canoe is to the island zoo.

The fourth and final consideration is developed fully in chapter 10. I will only outline it here. Whatever types of property electronuclear objects have, questions arise concerning how properties of those types are related to each other. If the remote quadrant is empty, the relational questions concern only three quadrants—the outward quadrant, the observable experiential quadrant, and the unobservable experiential quadrant. But if the remote quadrant is not empty, the relational questions concern all four quadrants. This makes the questions and the possible answers to them significantly different. In particular, it makes possible the hypothesis that *experiential properties are joint products of outwardly observable properties and remote properties*. I call this *the two-quadrant foundation hypothesis*. Chapter 10 describes this hypothesis in detail and explains why it provides a highly plausible account of the generation of animal experience.

Keep in mind that these four considerations are not four standalone arguments. They have their full force when considered as a group.

I have presented these considerations in support of the hypothesis that at least one wholly nonexperiential electronuclear object has at least one remote property. The remote quadrant is not empty. However, because these considerations are completely general, drawing no distinctions between different types of electronuclear object, they support the stronger hypothesis that remote properties are widespread if not universally present among electronuclear objects. This includes the electrons, protons, and neutrons that are the fundamental components of all larger electronuclear objects. It also includes electronuclear systems that have experiential properties. It is a basic tenet of the quadrants view that many and perhaps all electronuclear objects have remote properties. There is room for doubt about this, but the considerations just presented show that it is a plausible conjecture.

<>

If electrons and/or protons have remote properties, then all electronuclear objects, including all living things, have at least the *fundamental* remote properties of these fundamental components. What about *evolved* remote properties? Due to the extreme cognitive remoteness of remote properties, it is impossible to say with any confidence whether living things have evolved remote properties. However, one thing we can say is that there is no reason to rule out the evolution of new remote properties. Evolution is the evolution of plenary systems, so any quadrant could contain properties that are products of the evolutionary process. There is no reason to think that the evolution of new remote properties is any less possible than the evolution of new outwardly observable properties or the evolution of new experiential properties—both of which have occurred abundantly in the evolutionary process.

I conclude this chapter with two speculations about roles that evolved remote properties might play in animal life.

First, the dawn of experience at some unknown stage of animal evolution could have taken the form of a gradual transition from certain "advanced" remote properties to the simplest unobservable experiential properties. The transition could have gone through any number of quasi-experiential phases, for which we have no vocabulary. Some writers assume that experience first came into existence in an abrupt and discontinuous manner. Some writers reject the idea of an abrupt beginning of experience and argue for panexperientialism on the ground that it is the only other possibility. These are not the only possibilities. We are familiar with many gradual transitions that yield significant cumulative change. Some examples are the evolutionary transition from fish to land animals, the ripening of a fruit, and the education of a human being. If the fundamental components of electronuclear fabric have fundamental remote properties, then one can imagine a dawn of experience that was as subtly continuous as the dawn of a new day, involving a gradual transition from remote properties through quasi-experiential properties to simple experiential properties.

Second, we sometimes have vivid sensory recollections, such as a visual memory of a scene from childhood or an auditory memory of a familiar voice. For such a sensory recollection to be possible, something must be continuously stored in the brain that was produced by the original experience and that in turn gives rise to the recollection experience. In what format are such sensory memories stored? I find it plausible that they are stored in a format that consists in part of evolved remote properties. If sensory memories are stored in a partially remote format, then the storage format can resemble both the original experience and the recollection experience more closely than it can if the storage format consists entirely of outwardly observable properties. Closer resemblance means less conversion work—less work to produce the stored memory from the original experience and less work to produce a recollection experience from the stored memory. One might expect the brain to do what it does efficiently, and processes that link experiential properties to remote properties could be highly efficient. There is no conflict between this speculation and research that shows that the formation of new memories involves the formation and strengthening of synapses between nerve cells. The formation of a new memory is a change in plenary brain tissue. Therefore it could involve a coordinated combination of outwardly observable changes and remote changes.

8

Outwardism

The four quadrants differ in their cognitive distance from us, in the ways that are explained in the previous chapter. As a consequence, people tend to form skewed conceptions of the world that are unduly dominated by the cognitively closer quadrants. "Outwardism" is my term for any conception of the world that is dominated by the outward quadrant to a degree that there is good reason to believe the world itself is not. Such a conception involves an overweighting of the outward quadrant and an underweighting of some or all of the other quadrants, relative to the most reasonable assessment of each quadrant's actual share of plenary reality.

I believe that almost everyone is in the grip of outwardism. I include my former self. Recognizing and overcoming my former outwardism is a transformation that I went through rather late in life. Following are four factors that help to explain why outwardism has such a hold on people.

First, outwardly observable properties of objects are always in your face, so to speak. These are the properties that the objects advertise to us, and advertising works.

Second, the way we think about electronuclear objects tends to mimic our outward observational perspective on them. For example, we visualize objects with "the mind's eye." To think about an electronuclear object other than outwardly, you have to think about it without imagining it from an outward observational perspective. If you're like me, when you try to do this you'll experience a jolt of mental blankness. It's all but impossible. The perspective of outward observation is so deeply part of us, in both perception and thought, that we are ill-equipped to transcend it.

Third, we live in a society that cultivates and venerates the physical sciences. We are awash in descriptions of outward-observation-based scientific discoveries and theories. These accounts reinforce our everyday exposure to the more obvious outwardly observable properties of objects.

Fourth, the rich harvest of outward observation, especially vision with its many technological extensions, gives one a sense of cognitive abundance that does not encourage the thought that something is left out. There is outwardly obtainable data without end waiting to be collected and analyzed. Outward observation is a cognitive prison, but it is a gigantic prison whose locked doors are hidden in such a way that the prisoners do not feel confined.

In sum, our lives are replete with outward observation, thinking that mimics outward observation, prestigious reports of the outward-observation-based discoveries of physical scientists, and a sense of outwardly observable abundance. The result is widespread and deep-seated outwardism.

<>

One form of outwardism is what I call everyday outwardism. Consider all the electronuclear objects that are not living animals—plants, pebbles, and pencils, for example. In chapter 7 I mentioned the view known as panexperientialism, according to which experiential properties pervade the electronuclear fabric of such objects. There are people who believe in panexperientialism. There are more people who have heard of panexperientialism and reject it. But most people are in a third category: they have never entertained the thought that experiential properties might pervade the electronuclear fabric of all objects. They simply believe that some things, namely living animals of various kinds, have experience.

What about remote properties, as defined in chapter 7? It is not a widely held belief that all electronuclear objects have remote properties. But neither is it a widely held belief that electronuclear objects do not have remote properties. No belief regarding remote properties is widely held because the very idea of a remote property is unfamiliar to most people. People don't normally think about objects in this way.

With experiential properties and remote properties off the table, only one quadrant remains—the outward quadrant. And so we arrive at an interesting conclusion. Most people think about the nonliving objects in their environment entirely in terms of outwardly observable properties—so much so that they give no thought to whether these objects might have properties that are not outwardly observable. Whatever the plenary nature of plants, pebbles, and pencils might be, for most people such objects have a purely outward reputation.

People generally deal with these objects *as if* their plenary nature consisted only of outwardly observable properties, but without explicitly believing that this is the case. They draw no distinction between what these objects are and what properties these objects make available to observers through the spatial transmissions. They embrace the outward profiles of electronuclear objects as stand-ins for the plenary objects. They do not consider that the plenary objects might be propertywise plumper than their outward profiles. This is the psychology of everyday outwardism.

In his geometrical fantasy *Flatland*, Edwin A. Abbott (2005) describes a society of intelligent beings who live in a paper-thin domain that is embedded in three-dimensional space. The Flatlanders are paper-thin wafer-people who have the shapes of regular plane figures. The narrator, for example, has the shape of a square. I see an interesting parallel between the psychology of everyday outwardism and the "flat" psychology of the residents of Flatland. The senses of the Flatlanders are confined to their paper-thin domain, and so are their thoughts. Flatlanders do not believe that Flatland is embedded in three-dimensional space, but they don't disbelieve this either. They are dead to the possibility. They simply live life flatly. Analogously, the senses of human beings are dependent on certain types of spatial transmission, and people deal with nonliving objects in terms of what these spatial transmissions convey. In general, people neither believe nor disbelieve that these objects have properties of a kind that cannot be evidenced in the spatial transmissions. They are dead to the possibility. They simply relate to these objects outwardly. Thoughts about the possibility of a full-blown third spatial dimension are foreign to the fictitious Flatlanders. In like manner, thoughts about the possibility of remote properties in electronuclear objects are foreign to most real people.

Abbott, incidentally, uses his flatness fantasy to develop a different idea. He asks whether space might have additional dimensions that escape human beings, much as the third dimension escapes the Flatlanders. This is a fair question, but it is not a question that plays a role in this book. My case for the quadrants view remains the same whether or not the three-dimensional vast expanse is embedded in an expanse of more dimensions. The analogy that I am noting here is between Flatlanders being dead to the possibility that the world might be spatially plumper than their paper-thin domain and real people being dead to the possibility that everyday objects might be propertywise plumper than outward observation can find evidence for.

There is also an interesting parallel between everyday outwardism and the cognitive phenomenon that psychologists call *availability bias*. Experiments show that people often base their answers to general questions entirely on relevant examples that pop effortlessly into their thoughts. It tends not to occur to people that the examples they think of so easily might not be representative of the full set of relevant examples. For instance, when asked about the relative frequencies of different causes of death, people tend to overestimate the frequency of deaths due to accidents and tornados and underestimate the frequency of deaths due to strokes, asthma, and diabetes (Kahneman 2011, 138). The presumptive explanation is that deaths due to accidents and tornados receive more frequent and more graphic media coverage. Deliberate efforts by organizations to control what you hear about them, known euphemistically as "public relations," can have the same effect. So can the tendency of individuals to tout what they're proud of while keeping less flattering personal facts to themselves. Kahneman sums up the phenomenon as follows:

> The world in our heads is not a precise replica of reality; our expectations about the frequency of events are distorted by the prevalence and emotional intensity of the messages to which we are exposed. (Kahneman 2011, 138)

Analogously, our sense of the nature of electronuclear objects is distorted by the prevalence of the "messages" that these objects send us. Outwardly observable properties, being the subject of many "messages," are prominent in our thinking. Remote properties, being the subject of no "messages," are absent from our thinking. The result is an outwardly skewed conception of electronuclear objects.

In general, there are two ways to guard against availability bias. One way is to be aware of the existence of the bias-generating mechanisms. News media select stories for their entertainment value, organizations and individuals try to create positive impressions, and so on. The other way is to make a point of seeking out harder-to-find examples instead of relying lazily on the first examples that come to mind. However, when it comes to overcoming everyday outwardism, only the first of these methods applies. All you can do is appreciate the force of the plausibility arguments that electronuclear objects have remote properties in addition to their outwardly observable properties. You cannot seek out specific examples of remote properties in order to develop a more balanced conception of the objects, because remote properties are by definition impossible to find. Accordingly, it is psychologically more difficult to overcome everyday outwardism than to overcome the kind of availability bias that psychologists study; the general arguments for the quadrants view must carry the whole burden.

Leading the life of a physical scientist tends to reinforce the psychology of everyday outwardism. Research in the physical sciences brings to light more and more outwardly observable properties. Microscopes, telescopes, high-tech measuring instruments, and computerized imaging devices enlarge the set of outwardly observable properties that come to the attention of human beings. For someone who pursues a career as a physicist or a chemist, the objects that they study come to have increasingly complex but still purely outward reputations. This does not mean that a physical scientist must inevitably have an outwardly skewed conception of the world. Like everyone else, physical scientists are free to ponder the relationship between research in the physical sciences and the plenary nature of things, and they are free to move away from outwardism if that is where their thinking takes them. By itself, however, a career of research in the physical sciences tends to strengthen the grip of everyday outwardism.

<>

The plot thickens when the focus shifts from plants, pebbles, and pencils to human beings and their brains. Here, an outwardly skewed thinker cannot simply ignore the easy-to-ignore possibility of remote properties. He has to say something about the hard-to-ignore inwardly observable features of his own experience. Outwardly skewed thinkers have found a variety of things to say on this subject that enable them to persist in their outwardism.

There is a large group of writers who combine a purely outward conception of all electronuclear objects with the belief (which I think is correct) that experience is real and not outwardly observable. This combination leads to a body-plus view of human beings: experience is something "over and above" brain and body. Descartes had a purely outward conception of "extended things," so to give experience a home he posited unextended "thinking things." T. H. Huxley had a purely outward conception of the brain, so to give experience a home he posited "mental steam" emitted by the brain. The following passage by the linguist Derek Bickerton suggests this step from a purely outward conception of the brain to a body-plus view of human beings, but without explicitly taking it:

> There are no images in the brain. There are no words in the
> brain. All that's there are neurons and their connections and

> differential rates of electrochemical impulses. These provide
> a subjective sense of words and images. (Bickerton 2009, 194)

Bickerton does not explain how "a subjective sense of words and images" can be "provided" by the brain yet not be in the brain, but a serious attempt to explain this would seem to require a body-plus view.

There is a second large group of writers who combine a purely outward conception of all electronuclear objects with the belief (which I think is correct) that a human being is a human body. This combination leads them to say one of two surprising things. Some say that inwardly observable features of experience are also available to outward observation, while others say that there are no inwardly observable features of experience. Both of these views have found expression in so-called identity theories, which I discuss in chapters 20 and 21.

Most writers reserve the term "materialism" for this second group of views. This usage is itself a manifestation of outwardism. You might think that any body-only view of human beings would qualify as a form of materialism. For example, according to the quadrants view a human being is made entirely of electronuclear fabric. Electrons, protons, and neutrons are generally considered to be forms of matter. So isn't the quadrants view a form of materialism? It would be a mistake to answer with a simple yes, because the word "materialism" is not normally used this way. In its normal usage, the word "materialism" is tied to the claim that the outward-observation-based physical sciences can give us a complete understanding of human beings. Several examples of this usage follow.

D. M. Armstrong writes:

> A pure Materialist allows man nothing but physical, chemical
> and biological properties which, in all probability, he regards
> as reducible to physical properties only. (Armstrong 1968, 37)

The "physical, chemical and biological properties" that Armstrong refers to here are properties that figure in the outward-observation-based sciences of physics, chemistry, and biology. Accordingly, Armstrong's materialism is not just the claim that human beings are made entirely of matter; it has an additional component concerning the cognitive power of the outward-observation-based physical sciences.

Fred Dretske writes:

> For a materialist there are no facts that are accessible to only
> one person. (Dretske 1995, 65)

The claim here is that whatever one person can access through inward observation—if anything—others can access through outward observation. Dretske's materialism, like Armstrong's, combines the claim that human beings are made entirely of matter with a claim concerning the cognitive power of the outward-observation-based physical sciences.

Keith Campbell writes:

> One more step is required to reach Central-State Materialism. This step insists ... that the only properties the nervous system has are the properties recognized in chemistry and physics, together with their derivatives. Without this step the doctrine is not a materialism but a theory which accords to the brain two different sorts of attributes, non-material as well as material ones. (Campbell 1984, 87)

For Campbell, attributes of the brain are not "material" unless they are "recognized in chemistry and physics." It is not enough to be attributes of an organ that is made entirely of matter.

Hilary Putnam (who does not consider himself a materialist) writes:

> For a materialist philosopher, the 'factual' component in the meaning of any statement has to consist in a statement expressible in the vocabulary of *physical science*. (Putnam 1981, 204; his emphasis)

To be a "materialist philosopher," according to Putnam, you have to think that every fact about human beings is "expressible in the vocabulary of physical science." It is not enough to believe that human beings are made entirely of matter.

Paul Churchland coins the term "methodological materialism":

> The conviction of methodological materialism is that if we set about to understand the physical, chemical, electrical, and developmental behavior of neurons, and the ways in which they exert control over one another and over behavior, then we will be on our way toward understanding everything there is to know about natural intelligence. (Churchland 1988, 97)

A better name for this view would be "methodological outwardism" because its central idea is that human beings can be thoroughly understood using only the methods of the outward-observation-based physical sciences. There is no logical connection between this view and the belief that human beings are made entirely of matter.

Daniel Dennett writes:

> I propose to see, then, just what the mind looks like from the third-person, materialistic perspective of contemporary science. (Dennett 1987, 7)

Like me, Dennett believes that a human being is made entirely of matter. From this it follows that outward observation and inward observation are both perspectives on matter. If a university had a Department of Matter, both outward observation and inward observation would be relevant to its work. Why then does Dennett use the label "materialistic" for the "third-person" perspective of outward observation but not for the "first-person" perspective of inward observation? The answer is that in normal usage the term "materialistic" is not simply about matter. It is outwardly skewed, being tied to outward observation and the physical sciences.

According to the quadrants view, human beings are made entirely of matter, but the properties of objects that are made entirely of matter are not limited to properties that figure in the outward-observation-based physical sciences. They can include properties that are inwardly observable by only one person, as well as properties that are not observable at all—remote properties and the experiential properties of animals that have no capacity for inward observation. Therefore, the quadrants view is not a form of materialism in the usual sense of the word.

How did the word "materialism," which is built from the root "matter," get tied to these claims about outward observation and the cognitive power of the physical sciences? The answer, I believe, is that many writers have had a purely outward conception of matter, which has consequently gotten rolled into the meaning of the word "materialism." To be a "materialist" in the usual sense of the word, one has to believe that a human being is made entirely of matter *and also* that things made entirely of matter have only properties that can be accessed using the outward-observation-based methods of the physical sciences.

One writer who does not use the word "materialism" in this outwardly skewed way is Galen Strawson. In his essay "Real Materialism" (Strawson 2008, 19–51) he gives the word a definition that does not mention the physical sciences. Oddly enough, Strawson's definition does not mention matter either. I find it extremely vague, but for that very reason it seems broad enough to accommodate the quadrants view. Accordingly, it would be a mistake to say that the quadrants view is *not* a form of materialism. The quadrants view is or is not a form of materialism, depending on whose definition of "materialism" you are using. The fact that the word "materialism" is ambiguous in such a significant way is a good reason not to use it.

Similar considerations apply to the phrase "material object." If this phrase meant simply an object made entirely of matter, then I could say that a human being is a material object, or at any rate a material system. But I do not say this, because many writers roll an outward conception of matter into the meaning of "material object." In this usage, material objects are objects that have only outwardly observable properties. According to the quadrants view, human beings are not material objects in this sense. In fact, according to the quadrants view, there are no material objects in this sense. This is not because the things that are commonly called material objects do not exist; it is because the things that are commonly called material objects are electronuclear objects that have remote properties.

The writers in a third group hold body-only views and regard animals that have experience as exceptions to pure outwardism. These writers combine the belief that experience is not outwardly observable (which I think is correct), the belief that a human being is a human body (which I think is correct), and a purely outward conception of all *nonexperiential* electronuclear objects. They relax their outwardism enough to allow certain electronuclear objects to have experiential properties that are not outwardly observable, but they retain outwardism for the vast majority of electronuclear objects, which have no experience. A prominent member of this group is John Searle, whose view I examine in detail in chapters 13 and 22.

Some writers vacillate in a confusing way between statements that would put them in the first, body-plus group and statements that would put them in the third, body-only group. The vacillation seems to be the result of trying to combine three beliefs that are not logically compatible—the belief that experience is not outwardly observable (which I think is correct), the belief that a human being is a human body (which I think is correct), and a purely outward conception of the human body. One writer who vacillates in this way is Jaegwon Kim. Phrases such as the following suggest a body-plus view not too different from Huxley's:

> the apparent "gap" between phenomenal consciousness and the brain (Kim 2006, 220)
>
> the gap between consciousness and the brain (Kim 2005, 29)
>
> the explanatory gap between consciousness and the brain (Kim 2005, 121)

According to the quadrants view, the so-called gap that Kim refers to here is between the experiential properties of the brain and the outwardly observable properties of the brain. Presumably, Kim locates this gap between the brain and something else because he thinks of the brain in a purely outward way. Yet the following passage suggests that Kim does not do this wholeheartedly:

> Prima facie, an unbridgeable gap seems to exist, separating even the most complete and perfect knowledge of the brain as a biological/physical system from knowledge of the conscious experiences that may be going on in that brain. (Kim 2005, 94)

This murky sentence can be read as locating the gap between two kinds of knowledge of the brain, which are based respectively on outward observation and inward observation. This suggests a body-only view not too different from Searle's. To give another example, Kim makes the seemingly body-plus statement that "what happens in our mental life is wholly dependent on, and determined by, what happens with our bodily processes" (2005, 14), but he also makes the seemingly body-only statement that "Consciousness and intentionality are properties of biological organisms, or at least their neural systems" (1998, 83). Kim vacillates in this way throughout his many books and articles on the subject. Sometimes he casts experience as an aspect of the body, sometimes as an accompaniment of the body.

<>

In the frequently quoted paragraph 17 of *The Monadology*, Leibniz notes that one cannot visually discover experience "in a compound or in a machine," and from that fact alone he jumps to the conclusion that experience cannot be there:

> And supposing there were a machine, so constructed as to think, feel, and have perception, it might be conceived as increased in size, while keeping the same proportions, so that one might go into it as into a mill. That being so, we should, on examining its interior, find only parts which work upon one another, and never anything by which to explain a perception. Thus it is in a simple substance, and not in a compound or in a machine, that perception must be sought for. (Leibniz 1971, 228)

This is the epitome of outwardism—assuming that an outwardly observable object has only outwardly observable properties.

In a kindred passage, Fred Dretske describes the impossibility of visually discovering a person's experience in his brain:

> ... we all know what we will find when we look inside the skull. If there are sensations and feelings in there, they seem to be well camouflaged. When Wilder Penfield looked into the brain of his patient, a patient who was (as a result of electrical stimulation by Penfield) experiencing a variety of sensations, he did not observe the colors and sounds the patient reported experiencing.... what we find by looking in the brain of a person experiencing blue dogs is neither blue nor doglike. We do not find the content of experience, the properties that make the experience the kind of experience it is. What we find, instead, are electrical and chemical activity in gray, cheesy brain matter. (Dretske 1995, 36)

From this, Dretske jumps to the conclusion that "the experience-content isn't there" (1995, 37) and "the mind isn't in the head" (1995, 38). Again we have the epitome of outwardism—assuming that an outwardly observable object has only outwardly observable properties.

Leibniz and Dretske hold very different views. Leibniz, like Descartes, locates experience in "a simple substance" that is distinct from the body.

Dretske advocates an "externalist" theory of experience content that I am unable to make sense of. But both are led to deny experiential properties to the body by the same outwardly skewed principle: an outwardly observable object has only outwardly observable properties.

Thomas Nagel has written extensively in opposition to extreme forms of outwardism, but there is outwardism in his view too. Consider the following passage:

> Yet subjective aspects of the mental can be apprehended only from the point of view of the creature itself (perhaps taken up by someone else), whereas what is physical is simply there, and can be externally apprehended from more than one point of view. (Nagel 1979b, 201)

What is noteworthy here is the linkage of "is simply there" to "can be externally apprehended from more than one point of view," as if these two conditions go hand in hand. Because subjective aspects of the mental *cannot* be externally apprehended from more than one point of view, Nagel seems to be saying, they are not simply there. This is an unwarranted inference. It is logically possible for an object to have properties that are simply there yet cannot be externally apprehended. Failure to acknowledge this possibility is a symptom of outwardism. According to the quadrants view, moreover, this logical possibility is a ubiquitous reality. All remote properties and all experiential properties satisfy both conditions; they are simply there, but they cannot be outwardly observed.

If Nagel does not think that an animal's experiential properties are simply there, what does he think about them? His answer seems to be this:

> But the properties that make them experience exist only from the point of view of the types of beings who have them. (Nagel 1979b, 213)

If I am the judge, this sentence makes no sense. I can understand how something can be observable from only one point of view. I can understand how something can play a role in constituting a point of view. But I can think of no meaning for the phrase "exist only from a point of view." Nagel's outwardism leads him to deny the fully fledged existence of experiential properties and to relegate them to the obscure status of "existence from a point of view."

The following passage also reflects this way of thinking:

> ... it appears to be part of our idea of the physical world that what goes on in it can be apprehended not just from one point of view but from indefinitely many, *because* its objective nature is external to any point of view taken toward it ... (Nagel 1995a, 24; my emphasis)

In using the word "because" here, Nagel seems to be saying that the ability to be apprehended from many points of view *follows from* being external to all points of view. This is not the case, however, due to the logical possibility of properties that cannot be apprehended from any point of view. As a matter of simple logic, there are three possibilities—1) open to being apprehended from many points of view, 2) open to being apprehended from only one point of view, and 3) not open to being apprehended from any point of view. It is a symptom of Nagel's outwardism that he overlooks the third possibility. In chapter 13 I describe some similar reasoning by Searle.

Recall that I criticized Nagel in chapter 4 for seeming to claim that bats have a comprehensive point of view on their own experience. That claim is perhaps also a result of overlooking the third logical possibility. Once you grant that it is logically possible that the experiential properties of bats and many other animals cannot be apprehended from any point of view, it is hard not to take seriously the idea that this is in fact the case.

<>

I noted above that the terms "materialism" and "material object" are usually used with outwardly skewed meanings. The same is true of many other familiar terms. Examples follow.

The purely outward conception of the brain that leads many writers to embrace body-plus views and many others to embrace outwardly skewed body-only views is accompanied by an outwardly skewed usage of the word "brain." For example, it is common for writers to contrast "conscious states" with "brain states" in a way that suggests that a "brain state" can have only outwardly observable properties. If one uses the phrase "brain state" in a plenary sense, the contrast in question is between two aspects of a brain state—an aspect that consists of experiential properties and an aspect that consist of outwardly observable properties. Not all plenary brain states have an experiential aspect, but many do. "Brain state" in the outwardly skewed sense is limited to the scope of outward observation. "Brain state" in the plenary sense can include whatever properties the three-pound organ in the skull can have, without regard to how or whether people can observe them.

Discussing outward observation and inward observation in his book *Understanding Consciousness*, Max Velmans (2000, 249) poses the question "What is the one thing of which we have two, complementary forms of knowledge?" This is an important question and nicely framed too, but the answer Velmans gives reflects outward skewing. His answer is not "the human brain" or "the human body," but rather "the nature of mind." He goes on to explain this answer as follows:

+   In the human case, minds viewed from the outside seem to take the form of brains (or some physical aspect of brains).
+   Viewed from the perspective of those who embody them, minds take the form of conscious experiences. (Velmans 2000, 249–250)

Here, seemingly, an outwardly skewed usage of the word "brain" has led Velmans to invent an object—which he calls "the mind" or "the nature of mind"—whose outwardly observably aspect is the brain. This is backwards. A lung or a kidney is an object, not an aspect of some object other than itself. Likewise, a brain is not an aspect of a mind-object; a brain is an object having certain aspects, some of which we subsume under the word "mind." A living plenary brain has properties of various types; some are outwardly observable, some are inwardly observable, and some are neither.

Many writers discuss the "neural correlates of consciousness." Here the correlations in question are between experiential properties and *outwardly observable* neural properties; the adjective "neural" is implicitly restricted to outwardly observable properties of the nervous system. The same outward skewing of "neural" is suggested by a number of other expressions including "the neural basis of consciousness," "the neural substrate of consciousness," and "the neural underpinnings of consciousness." One sees it in Jaegwon Kim's contrast of "mental" with "neural" when he says that "the causal role of *a mental property* had by me is threatened with preemption by another property, *a neural property*, also had by me" (1998, 117; my emphasis). If one uses the adjective "neural" in a plenary sense, meaning simply "pertaining to the nervous system," it does not contrast with but rather *subsumes* "conscious" and "mental." Assuming the correctness of a body-only view, all conscious properties and all mental properties are neural properties in this plenary sense. Using "neural" in this plenary sense, the so-called "neural correlates of consciousness" are outwardly observable neural properties that are correlated with inwardly observable neural properties. "Neural" in the outwardly skewed sense is limited to the scope of outward observation; "neural" in the plenary sense includes whatever properties the nervous system has, without regard to how or whether people can observe them.

Another important word that has both plenary and outwardly skewed senses is "biological." In my dictionary, the first definition of this adjective is "of or relating to biology or to life and living processes." This definition throws together two significantly different ideas as if there were no significant difference between them. If "biology" means *the science of biology,* there is a crucial difference between "of or relating to biology" and "of or relating to life and living processes." The reason is that the plenary reality of life and living processes includes experiential properties and possibly also remote properties, which do not figure in the outward-observation-based science of biology. To treat these two senses of "biological" as one is to tacitly equate the scope of the outward-observation-based science of biology with the plenary reality of life and living processes. The result is outwardism regarding the nature of life.

The following sentence is interesting because it uses the word "biological" twice in quick succession, shifting from one meaning to the other in the process:

> The mind may be a biological product, but biological concepts
> can provide us with only a superficial understanding of its
> contents and workings. (Nagel 1995b, 37)

If one tries to understand these two occurrences of "biological" in the same way, it is hard to see how this sentence could express a reasonable thought. Yet it does express a reasonable thought. The reason is that the mind is a "biological product" in the sense that it is a product of the plenary evolution of living things, whereas "biological concepts" are concepts that figure in the outward-observation-based science of biology. Nagel's paradoxical statement is true because the outward-observation-based science of biology deals with only the outward profile of plenary living things.

The same tendency to outward skewing taints the word "evolution." In *The Making of Memory,* Steven Rose describes the evolution of living things in the following way:

> Evolution, which literally means unrolling, or development,
> is, for biologists, the process by which there has occurred a
> steady change in the form of organisms across generations.
> By form here is meant anything from their biochemistry and
> internal structure to their behavior. (Rose 1992, 159)

Why does Rose say "a steady change in the *form* of organisms" instead of "a steady change in the *characteristics* of organisms" or simply "a steady change in organisms"? Why does he limit his definition of evolution to outwardly observable properties? Presumably, the answer lies in the little phrase "for biologists." Biologists work in an outward-observation-based scientific discipline, and so "for biologists" evolution concerns only the outwardly observable properties that they study. By inserting this qualification, Rose lowers the dark curtain of outwardism. Biologists study the evolution of living things in the ways that they can. This is as it must be. But there is no justification for narrowing the meaning of the word "evolution" so that it includes only what biologists can study. Rose offers an outwardly skewed definition of "evolution" that excludes three of the four quadrants of the plenary evolution of living things.

The familiar scientific terminology of "genotype" and "phenotype" is also shaped by outwardism. A typical definition of "phenotype" is "the observable properties of an organism that are produced by the interaction of the genotype and the environment." Why this restriction to observable properties? What happened to the plenary organism, the organism with all its properties regardless of how or whether they can be observed? An actual living plant or animal exemplifies not just an outward-quadrant phenotype, but a multiple-quadrant *plenary-type* (my coinage). It is convenient for biologists to forget about the plenary organism because biology's outward-observation-based methods provide no access to three of its four quadrants. Focusing on the outward-quadrant phenotype makes biologists feel more knowledgeable about life than they really are.

Outward skewing taints several terms whose root is "nature." For example, the following passages describe one sense of the term "naturalism":

> As a preliminary characterization, we take naturalism
> to be the view that only natural science deserves full and
> unqualified credence. (Wagner and Warner 1993, 1)

> Naturalism, understood broadly as the view that the account of nature provided by the natural sciences is our only guide as to what genuinely, or unproblematically, exists and/or to what is genuinely, or unproblematically, known, is widely popular within contemporary analytic philosophy. (De Caro 2008, 107)

Both of these descriptions are vague, but one thing that is clear about them is that they tie the term "naturalism" to natural science. The result is outward skewing because the natural sciences are based entirely on outward observation.

The explanation of why "natural science" is limited in this way has several parts. First, remote properties and unobservable experiential properties get left out of natural science because they are not observable. Second, most of nature does not have inwardly observable experiential properties, so in general students of nature can go about their business without considering such properties, even though they are observable. And third, in days of old when the term "natural science" first gained currency, most people held the traditional religious view that inward observation of experience is observation of a nonbodily soul that is not even part of nature. Due to this combination of reasons, "natural science" is focused exclusively on the outward quadrant and thus leaves significant aspects of nature out of account.

The word "naturalism" is not always tied to the natural sciences in this way. Stephen Cave speaks of "naturalism—that is, a worldview that rejects the supernatural" (2012, 285). This is a broader definition, free of cognitive skewing. Several writers put an adjective in front of "naturalism" to construct a label that they find congenial. There is John Searle's "biological naturalism," Colin McGinn's "transcendental naturalism," Gregg Rosenberg's "liberal naturalism," and Christian de Quincey's "radical naturalism." The last two are panpsychist views, which are discussed in chapter 12. Searle's view is discussed in chapter 13. McGinn's view is discussed in chapters 14 and 15. All of these two-word phrases are less outwardly skewed than natural-science "naturalism."

There is an outwardly skewed, natural-science sense of the verb "naturalize" that one writer defines as follows:

> ... to "naturalize" means to explain without recourse to a mentalistic vocabulary within a theory that can finally become part of the natural sciences. (Kurthen 1995, 107)

Suppose, plausibly, that there is a significant aspect of certain electronuclear systems that can *only* be discussed by using a mentalistic vocabulary. According to the definition given here, one can only "naturalize" such a system by disregarding much of its nature!

There is an interesting illustration of the outward skewing of "nature" terms in the book *Aping Mankind* by Raymond Tallis. Tallis criticizes certain writers for "rejecting supernatural accounts of human nature only to embrace the opposite error of concluding that humans must therefore be simply parts of the natural world." He goes on to say:

> It is because I do not believe that rejecting a divine origin of the universe in general, or of us in particular, necessarily leads to a naturalistic account of what we are that I have written this book. (Tallis 2011, 327–328)

In the same vein, Tallis writes:

> It does not seem to me a very great advance to escape from the prison of false supernatural thought only to land in the prison of a naturalistic understanding. (Tallis 2011, 10)

These sentences sound illogical. Rejecting the supernatural leaves you with the natural, one might reasonably think; there is nowhere else to go. However, this is not the case if the words "natural" and "naturalistic" are used in an outwardly skewed way. And this seems to be how Tallis uses them. He says that he opposes "a tendency to overestimate what natural science, in particular biology, has to say about human nature" (2011, 3). What he seeks, stated in clearer terms than his, is an understanding of human beings that is neither supernatural nor limited to what outward-observation-based natural science has to offer. That is not illogical at all.

I must add that I have little sympathy for the way that Tallis pursues this reasonable goal. Instead of giving a substantive account, he uses cryptic phrases such as "the human world, created out of pooled transcendence" (2011, 314,) "a common space outside nature" (2011, 242), and "the community of minds, which has left the biosphere behind" (2011, 238). But I agree with him that human beings have properties that are neither supernatural nor dealt with by natural science. According to the quadrants view, not only human beings but all electronuclear objects have properties that are neither supernatural nor dealt with by natural science. Three of the four quadrants satisfy this condition.

The following sentence is interesting for the way it connects broad and narrow senses of the word "natural" as if there is no difference between them:

> The determinist standpoint is often linked to **scientific naturalism,** which is the position that all events are natural (as opposed to supernatural) and so explicable according to natural laws. (Bardon 2013, 144)

The use of "and so" in this sentence suggests that being "explicable according to natural laws" somehow follows from not being supernatural, yet in fact there is no logical connection here. It is logically possible that certain things are neither supernatural nor explicable according to natural laws. Indeed, to find plausible candidates one need look no further than the appalling assortment of depraved acts that pepper the daily news.

Materialism, brain, neural, biological, evolution, phenotype, naturalism, naturalize, natural—the widespread practice of using important words like these in outwardly skewed ways helps to tighten the grip of outwardism. One way to loosen that grip is to start using these words in a plenary sense that leaves room for all the properties of electronuclear objects, without regard to how or whether people can observe them.

9

The Quadrants and the Word "Physical"

The word "physical" is almost always used in discussions of this subject. In addition to whatever it means, it seems to have an agreeable effect on many who use it, like a trusted brand.

What does "physical" mean? To answer this question well, one must consider the word in the context of a variety of phrases. Accordingly, this chapter discusses many commonly used "physical" phrases. It does not discuss all of them—there are too many to make that practical—but it discusses enough of them to bring out certain important points.

Some "physical" phrases have fairly clear meanings. "Physical education" concerns the development of muscles and motor skills. "Physical fitness" and "physical therapy" also concern muscle use. "Physical appearance" is visual appearance, how someone or something looks. Being "physically attractive" also concerns visual appearance. The popular song line "Let's get physical" is about two people touching each other. "Physical contact" is also about touching. "A physical" is a medical examination, which typically involves a lot of looking and touching. "Physical proximity" means that the spatial separation between two things is not great. The phrase "the physical sciences," which I discussed briefly in chapter 3, is an umbrella term for fields of study that depend exclusively on outward observation, such as physics, chemistry, and astronomy. Fields of study in which inwardly encountered thoughts, feelings, and sensations play a role, such as psychology, linguistics, and economics, are not normally classified as physical sciences. The word "physical" does not seem to have a single meaning across all these phrases; it seems to be associated variously with muscles, vision, touch, and space. However, the meaning of each of these "physical" phrases as a whole is fairly clear.

<>

"Physical" phrases that have significant ambiguities are more common. Some of these ambiguities are especially significant if the quadrants view is true.

I discussed the ambiguity of the phrase "physical space" in chapter 3. Sometimes this phrase is used to refer to something independent of thought that has existed for billions of years. Sometimes it is used to refer to certain theoretical constructions of physicists and cosmologists. To avoid the misunderstandings that this ambiguity can produce, I chose not to use the phrase "physical space." Instead, I have been calling the thought-independent home of electronuclear objects *the vast expanse.*

The much-used phrases "physical property" and "physical object" are related in a way that makes it appropriate to discuss them as a pair.

One way to understand the phrase "physical property" is as a shortened form of "physical-science property." Physical properties in this sense are outwardly observable properties that figure in the fields of study that we call the physical sciences. One might narrow this definition to include only outwardly observable properties that figure in the field of physics. However, this introduces worries about whether a property is inside or outside the fuzzy boundary between physics and other physical sciences. Alternatively, one might broaden this definition and equate "physical properties" with "outwardly observable properties." This avoids worries about the fuzzy boundary between physics and other physical sciences as well as worries about the fuzzy boundary of science. An important characteristic that all three of these variants have in common is that the scope of "physical property" falls entirely within the outward quadrant. If "physical property" means something like "physical-science property," then any experiential properties or remote properties that are not outwardly observable are not physical properties.

Given this definition of "physical property," one can go on to define "physical object" in either of the following two ways.

One possibility is to define "physical object" as any object that can be studied by the outward-observation-based methods of the physical sciences. In other words, a physical object is any object that has *some* outwardly observable, physical-science properties. By this definition, all electronuclear objects are physical objects. If the quadrants view is correct, these definitions together entail that many if not all physical objects have some nonphysical properties. This is a linguistically awkward implication that one might wish to avoid.

A second possibility is to define "physical object" more narrowly as an object that has *only* outwardly observable, physical-science properties. This narrower meaning seems to be the one at work in the following passage:

> If materialism is true, then human beings are large collections
> of small physical objects and nothing more, ontologically.
> *It follows that* any human being could be described, and
> described completely, in purely scientific terms. (Lycan 1996,
> 45; my emphasis)

Here, the second claim follows logically from the first claim, as the author asserts, if "physical objects" have only physical-science properties, but it does not follow if "physical objects" can have other properties in addition to their physical-science properties. If the quadrants view is correct, this more restrictive definition of "physical object" has the linguistically awkward implication that there are no physical objects. There are lots of electronuclear objects, but if they all have remote properties, then no electronuclear object is a physical object in this more restrictive sense. Sticks, stones, and everything else that we are in the habit of calling physical objects fail this demanding test. This is the same point that I made in chapter 8 concerning the phrase "material object." We are surrounded by material, physical objects—provided that we don't mean too much when we say this. If we mean that we are surrounded by objects that have only outwardly observable, physical-science properties, we are saying something that there is no reason to believe and that according to the quadrants view is not true.

Instead of defining "physical property" first and then defining "physical object" in terms of it, one can proceed in the opposite direction. First define a physical object as any object that can be studied by the outward-observation-based methods of the physical sciences, and then define a physical property as any property of a physical object. If the quadrants view is correct, this pair of definitions has the implication that all the properties in all four quadrants are physical properties. It also has several linguistically awkward implications that one might wish to avoid. First, there are many physical properties that are not physical-science properties, namely all experiential properties and all remote properties. Second, one cannot contrast "physical" properties with "mental" properties because by this definition the class of mental properties is a subclass of the class of physical properties. Third, one must give up use of the term "psychophysical" because by this definition all psychological properties are physical properties. The contrast that the term "psychophysical" is ordinarily used to mark becomes the contrast between inwardly observable physical properties and outwardly observable physical properties.

To sum up, I have discussed two definitions of "physical property" and two definitions of "physical object," which can be paired with one another in three different ways. From this array of possible definitions two interesting conclusions emerge.

First, it is difficult if not impossible to find a way to use the phrases "physical object" and "physical property" that does not have a linguistically awkward implication. The following table summarizes the three possible pairs of definitions I have considered:

	Paired Definitions of "Physical Property" and "Physical Object"	Linguistically Awkward Implication
1	Physical properties are outwardly observable, physical-science properties. Physical objects are outwardly observable, physical-science objects (objects that have *some* physical-science properties).	Many if not all physical objects have some nonphysical properties.
2	Physical properties are outwardly observable, physical-science properties. Physical objects are objects that have *only* physical properties.	There are no physical objects.
3	Physical objects are outwardly observable, physical-science objects. Physical properties are properties of physical objects.	Many physical properties are not physical-science properties. All mental or psychological properties are physical properties.

The root of this difficulty is the variety of properties that electronuclear objects can have, according to the quadrants view. The situation is too complex to be well described using these popular "physical" phrases. The popularity of the phrases goes hand in hand with insensitivity to the complexity of the situation.

Second, the differences between the definitions are extremely consequential. If "physical property" is equivalent to "physical-science property," as in pairs 1 and 2, then only properties in the outward quadrant are physical properties. If "physical property" covers all properties of physical-science objects, as in pair 3, then all properties in all quadrants are physical properties. The classification of the properties in three of the four quadrants hangs in the balance! Similarly, if "physical object" is equivalent to "physical-science object," as in pairs 1 and 3, then the vast expanse contains innumerable physical objects, but if "physical object" means an object that has only physical-science properties, as in pair 2, then there are no physical objects. The classification of every electronuclear object hangs in the balance!

Writers have defined "physical property" and "physical object" in a variety of other ways, which further illustrate the indefiniteness and ambiguity of these phrases. Following are three examples, which are by no means exhaustive.

First, Daniel Stoljar suggests that "physical property" can be defined in the following way:

> ... a physical property is a property which *either* is the sort of property that physical theory tells us about *or* else is a property which metaphysically (or logically) supervenes on the sort of property that physical theory tells us about. (Stoljar 2004; his emphasis)

This statement passes the buck from "physical property" to "physical theory." What is a "theory" and what distinguishes a "physical theory" from other theories? In a footnote, Stoljar hints at several ways in which "physical theory" might be understood, thereby turning this apparently single definition into a family of definitions. The absence of any characterization of the "sort" of property involved adds further ambiguity.

Second, taking a different approach, Stoljar introduces the phrase "paradigmatic physical object" in the following way:

> First, we point out that we have an ordinary conception of what a paradigmatic physical object is; that is, stones and

trees are paradigmatic physical objects, and numbers and propositions are not. (Stoljar 2006, 30)

This is inadequate because such a small set of examples does not suffice to define a "conception" or even to show that there is a conception. People do use the phrase "physical object" a lot, but is this usage backed by a conception or is it just chatter? Furthermore, Stoljar does not discuss the relation between "paradigmatic physical object" and "physical object." Are there physical objects that are not paradigmatic physical objects, in his terminology? If so, what is the basis of that distinction? And if not, why bother with the qualifier "paradigmatic"?

Third and last, Frank Jackson makes the following proposal:

> For, first, physicalists can give an ostensive definition of what they mean by physical properties and relations by pointing to some exemplars of non-sentient objects—tables, chairs, mountains, and the like—and then say that by physical properties and relations, they mean the kinds of properties and relations needed to give a complete account of things like them. (Jackson 1998, 7)

The big problem with this definition is the expression "give a complete account." What does it mean? Taking it in a straightforward way, there is no reason to believe that one *can* give a complete account of any electronuclear object, and some reason to suspect that one cannot. According to the quadrants view, one cannot give a complete account of any electronuclear object because one cannot discuss its remote properties except as an undifferentiated group. It is doubtful whether one can even give a complete account of an electronuclear object's outwardly observable properties; maybe the world is too complex to permit that. In passages such as the following, Jackson suggests that there is reason to believe that one can give complete accounts of electronuclear objects:

> ... we know that science can in principle tell us the whole truth about physical objects. (Jackson 1998, 2)

> ... it is reasonable to suppose that physical science, despite its known inadequacies, has advanced sufficiently for us to be confident of the kinds of properties and relations that are needed to give a complete account of non-sentient reality. (Jackson 1998, 7)

> It is implausible that there are facts about very simple organisms that cannot be deduced a priori from enough information about their physical nature and how they interact with their environment, physically described. The physical story about amoeba and their interactions with their environment is the whole story about amoeba. (Jackson 1998, 83)

But notice that he merely uses the reassuring words "know," "reasonable" and "implausible" without saying anything to justify their use. These passages give voice to a baseless outwardism.

<>

For anything that has properties, one can make the test of its being "physical" either 1) that *some* of its properties are physical-science properties or 2) that *all* of its properties are physical-science properties. Accordingly, there are many nouns that combine with the adjective "physical" to yield a phrase that is ambiguous in the same way that "physical object" is.

Consider next the phrase "physical particle." J. J. C. Smart writes:

> A man is a vast arrangement of physical particles, but there are not, over and above this, sensations or states of consciousness. There are just behavioral facts about this vast mechanism ...
> (Smart 1959, 54)

I can agree with Smart that a human being is a vast arrangement of particles, provided that "arrangement" is understood dynamically and "particle" is understood broadly. But are these particles physical particles? That depends on what one means by "physical particle." If a physical particle is a particle that has some physical-science properties, then yes, these electrons and protons and neutrons that we are made of are physical particles. But if a physical particle is a particle that has only physical-science properties, then there is no reason to think that we have any physical particles in us. All of the particles in us might well have remote properties. As I read Smart, he understands "physical particle" as a particle that has only physical-science properties, which makes his statement false according to the quadrants view. The particles that we are made of are not physical particles in Smart's sense.

Four other nouns that yield the same sort of ambiguity are "system," "process," "device," and "matter." Many of the illustrations that follow come from the writings of Paul Churchland, who uses the word "physical" extensively.

Churchland says that the brain is "a complex physical system" (1988, 2). What is a "physical system," and is the brain one? If a physical system is a system that has *some* physical-science properties, then the brain is obviously a physical system. But if a physical system is a system that has *only* physical-science properties, then there is no reason to believe that there is such a thing as a physical system anywhere in the world, and there is good reason to think that the brain is not one. A living brain has many transient experiential properties in addition to remote properties, and none of these are physical-science properties. The outward-observation-based physical sciences can tell us only about the brain's outward profile, not its plenary nature.

Churchland often qualifies the adjective "physical" with the adverb "purely." For example, he speaks of "the purely physical brain" (1995, 124). What does this mean? If it means an organ that has only physical-science properties, then the purely physical brain is a purely imaginary object.

David Chalmers uses the phrase "physical system" in the following passage:

> The hardest part of the mind-body problem is the question:
> how could a physical system give rise to conscious experience?
> (Chalmers 1996, 25)

An adequate response to this question must deal with the ambiguity of "physical system." If a physical system is a system that has *only* physical-science properties, then it is not the case that a physical system gives rise to consciousness. Puzzling over how a system of that sort gives rise to consciousness is misguided. On the other hand, if a physical system is a system that has *some* physical-science properties, then a physical system does give rise to consciousness but there is no way to understand how it does so because the process of generating experiential properties involves remote properties in addition to physical-science properties.

Churchland says that the development of a human embryo is a "purely physical process":

> The same theorists would also cite the now familiar fact that
> each individual person begins life as a sphere of interlocking
> protein molecules enclosing a cell nucleus filled with DNA
> molecules, and that he or she develops from there by a long

> and intricate *but purely physical* process. (Churchland 1995, 211; his emphasis)

What is the meaning of the phrase "purely physical process"? If it means a process all of whose participating objects have *some* physical-science properties, then the development of a human embryo is a purely physical process. But if it means a process all of whose participating objects have *only* physical-science properties, then there may not be any purely physical processes and in any case the development of a human embryo is not one. Understood in this way, Churchland's "now familiar fact" is a falsehood. From the outset, the development of a human embryo includes remote properties, and after a certain point it may well include experiential properties. None of these are physical-science properties. The outward-observation-based physical sciences can tell us only about the outward profile of a developing human embryo, not its plenary nature.

Churchland also describes the evolution of living things as a "purely physical process":

> For the purposes of our discussion, the important point about the standard evolutionary story is that the human species and all its features are the wholly physical outcome of a purely physical process. (Churchland 1988, 21)

Here it is worth recalling the outwardly skewed definition of "evolution" that I discussed in chapter 8. If Churchland's "standard evolutionary story" is an outwardly skewed story that mentions only properties that are available for outward-observation-based scientific study, then his statement may be correct, but only because the standard evolutionary story is outwardly skewed. The evolution of living things is not a process all of whose participating objects have *only* physical-science properties. The evolution of living things has included remote properties from the outset and experiential properties for quite some time. None of these are physical-science properties. The outward-observation-based physical sciences can tell us only about the outward profile of the evolution of living things, not its plenary nature.

Chalmers also poses his question using the phrase "physical process":

> The hard problem of consciousness ... is that of explaining how and why physical processes give rise to phenomenal consciousness. (Chalmers 2010, 105)

Again, one must deal with the ambiguous "physical" phrase in order to respond adequately to this question. If physical processes are processes all of whose participating objects have *only* physical-science properties, then it is not the case that physical processes give rise to phenomenal consciousness. Puzzling over how processes of that sort gives rise to consciousness is misguided. On the other hand, if physical processes are processes all of whose participating objects have *some* physical-science properties, then physical processes do give rise to phenomenal consciousness but there is no way to understand how they do so because the process of generating experiential properties involves remote properties in addition to physical-science properties.

Note that what Chalmers calls the hard problem of consciousness can be stated in plainer terms by eliminating the word "physical." The fundamental question is: how do the systems and processes that give rise to consciousness do so? This is not an isolated example. I have read many sentences by many writers that can be clarified by eliminating a superfluous occurrence of the adjective "physical."

Churchland (1988, 99) begins a chapter with the question, "Is it possible to construct and configure a purely physical device so that it will possess genuine intelligence?" Let's lop off the last seven words of this question and ask simply: Is it possible to construct and configure a purely physical device? If "purely physical device" means a device that has *some* physical-science properties, then the answer is obviously yes. But if "purely physical device" means a device that has *only* physical-science properties, then the answer is no. It is not possible to construct a purely physical device in this sense, because every device that we can construct is made of electronuclear fabric, which has remote properties. No electronuclear device is purely physical in this sense.

What about the phrase "physical matter" (Churchland 1988, 18, 99)? If this phrase means "matter that has *some* physical-science properties," then it is coextensive with the unqualified noun "matter." But if it means "matter that has *only* physical-science properties," then it refers to nothing, according to the quadrants view. Physical matter in this sense is mythical matter. Real matter has remote properties. The outward-observation-based physical sciences can tell us only about the outward profile of real matter, not its plenary nature.

<>

Jaegwon Kim makes extensive use of the phrase "the physical domain." In one book, he devotes six pages to a survey of the content of this "domain"

(1998, 112–118). He begins this survey by including "the basic particles and their properties and relations." To this he adds objects of many sizes:

> Plainly the physical domain must also include aggregates of basic particles, aggregates of these aggregates, and so on, without end; atoms, molecules, cells, tables, planets, computers, biological organisms, and all the rest must be, without question, part of the physical domain. (Kim 1998, 113)

Then he turns his attention to properties:

> What then of properties? What properties, in addition to the properties and relations of the basic particles, are to be allowed into the physical domain—that is, to be counted among physical properties? (Kim 1998, 113)

To answer this question, he runs through a series of rather complex conditions any one of which suffices to make a property a physical property. What is odd about this survey is that the resulting "physical domain" can include an object without including all of that object's properties. Such a domain seems strangely abstract and gerrymandered. Does it even make sense? In any case, there are less strange ways in which one could define a "physical domain" that contains both objects and properties. One could say that "the physical domain" contains 1) all the objects that Kim lists plus all the properties of those objects, or 2) all the properties that satisfy Kim's property conditions plus any objects that have only such properties. I see no reason to define "the physical domain" in the way that Kim does instead of in one of these simpler ways. Kim's definition seems not only strange but arbitrary. It is noteworthy that Kim's six-page survey of "the physical domain" does not mention space. However, Kim sometimes writes about "the physical domain" as if it includes the entire vast expanse. This is a further ambiguity of the phrase. All these ambiguities carry over to the principle that Kim calls "the causal closure of the physical domain," which I discuss in chapter 22.

Equally problematic is the phrase "the physical world." In his book *The Nature of the Physical World*, Arthur Eddington uses this phrase to refer to what is usually called a theory or a model. For example, there is a chapter entitled "World Building" in which he explains how "the physical world" has been "built" and "constructed" using certain abstract concepts and mathematical equations that interconnect the numerical "pointer readings" that scientists obtain with measuring instruments. The physical world in this sense is man-made and did not exist before the development of modern science. This usage is reminiscent of Feigl's use of the phrase "physical space" to mean not the vast expanse that contains the stars and planets but rather a set of thoughts that physicists have had about the vast expanse.

Eddington's usage is not common. More often, writers use the phrase "the physical world" to refer to something very large that does not owe its existence to human life and thought. Used in this way, the phrase "the physical world" has further ambiguities. It would seem that the physical world must be something less than the whole world; else why insert the adjective "physical"? Yet few users of this phrase bother to explain what part or aspect of the whole world they mean to exclude from "the physical world." It could be Cartesian thinking things, or "mental steam" produced by brains, or "nonphysical" properties defined in one way or another, or something else. There are many possibilities. Worse, there seem to be no good possibilities, because I see no good way to divide the whole world into the physical world and something else.

Roger Penrose concludes an essay on some abstruse topics with the following declaration:

> I cannot believe that it is impossible to make sense of the physical world. I think it is the task of physics to do precisely that. (Penrose 1986, 61)

This statement sounds good mainly because of the rhetorical pairing of "physical" with "physics." A casual reader might think "Physical world, physics—right, of course." But suppose we delete the adjective "physical," yielding the following:

> I cannot believe that it is impossible to make sense of the ...
> world. I think it is the task of physics to do precisely that.

This almost identical sentence makes a very different impression. One wonders, "Why assign the task of making sense of the world exclusively to physics, and what exactly is physics anyway?" The puzzle is whether there is a difference of meaning between "the world" and "the physical world" that can justify these sharply different reactions.

I find the phrases "the external physical world" (McKinsey 1996, 355) and "the external, physical world" (Velmans 2000, 106) intriguing because they suggest that these two writers use "the physical world" and "the external world" interchangeably. Accepting this equivalence would give the phrase "the physical world" all the problems of the phrase "the external world," which I discussed in chapter 5.

At times Kim seems to use "the physical world" interchangeably with "the physical domain," whose problems I have just discussed. On the other hand, Kim wrote a book titled *Mind in a Physical World*, which discusses the problem of "finding a place for the mind in a world that is fundamentally physical" (1998, 2). This problem description suggests that Kim's title might be somewhat loosely worded, letting the whole world qualify as "a physical world" by virtue of its being "fundamentally" physical. This would be like calling a ring "a gold ring" because it is made *mostly* of gold.

By the way, is Kim right that the world is fundamentally physical? It depends on what he means by "fundamentally physical"; this is yet another ambiguous "physical" phrase. If Kim means that all fundamental properties are physical-science properties, then he is not right, according to the quadrants view. The world is not fundamentally physical in this sense because there are fundamental remote properties. Undoubtedly we live in a world. Undoubtedly we live in a world that allows the outward-observation-based research of physical scientists to meet with considerable success. But do we live in a fundamentally physical world? That is an ambiguous question that does not have a straightforward answer.

Another ambiguous "physical" phrase is "the physical," in which the word "physical" is used in an adjectival manner but not followed by a noun. An example is the title of Herbert Feigl's book, *The "Mental" and the "Physical"*. Does "the physical" in this usage include objects or properties or both? Can it include objects without including all of their properties, like Kim's version of "the physical domain"? Users of the phrase "the physical" do not discuss these questions.

<>

Also of interest is the phrase "physical duplicate" and equivalents such as "physical replica," "exactly identical physically," and "wholly alike physically." These phrases have been used mainly to pose the question whether there could be a "physical duplicate" of you or me that is experientially blank. Some writers call such a hypothetical experientially blank humanoid a zombie—not to be confused with the different kind of zombie found in movies. There are two ways to understand the phrase "physical duplicate," and they yield different answers to this question, if the quadrants view is correct. On one understanding, a physical duplicate of me must be made of electrons, protons, and neutrons, just as I am, put together in exactly the same way and doing exactly the same things. Since I am an electronuclear system, such a "physical duplicate" of me would be a total duplicate of me. I am not experientially blank, so it could not be. On another understanding, a physical duplicate of me must merely be indistinguishable from me using the outward-observation-based methods of the physical sciences. Such a "physical duplicate" need not be made of electrons, protons, and neutrons. It could be made of electroids, protoids, and neutroids—hypothetical particles that are outwardly indistinguishable from electrons, protons, and neutrons, respectively, but which have different (or no) remote properties. Such a creature would not be a total duplicate of me, but only an outwardly indistinguishable imitation. Since it would lack the remote properties that are part of the causal basis of my experience, it could conceivably be experientially blank. I think it is enormously improbable that there are any such particles—exactly like electrons, protons, and neutrons in every way one can detect yet different in undetectable ways—but it is an idea that one can play with.

<>

There are many other ambiguous or obscure "physical" phrases in use, but I think I have surveyed enough of them to make the point that clarity is generally better served by avoiding the adjective "physical." The good news is that this is fairly easy to do. The fact that this adjective is used so much, and in combination with so many different nouns, tends to create the impression that it is indispensable for making certain points. But this is not the case. It's just a word with a positive aura whose use has gotten out of hand. In serious discussions of the nature of things, let's get less "physical."

<>

The ambiguities of the word "physical" carry over to the term "physicalism." Arguing for or against physicalism is a popular pursuit, but unfortunately the participants in this debate do not have a shared understanding of what issue they are discussing.

For many, physicalism is the view that physics, or physical science in general, is capable of providing a complete understanding of the world, or at least a basis from which a complete understanding of the world can be constructed. According to the quadrants view, physicalism in this sense is an extreme form of outwardism, a Flatland mentality that mistakes the world's outward profile for its plenary nature.

But the word "physicalism" has also been given definitions that might be compatible with the quadrants view. For example, John Searle writes:

> Suppose we define 'naïve physicalism' to be the view that all that exists in the world are physical particles with their properties and relations. (Searle 1984, 26–27)

This definition inherits the ambiguity of "physical particle" that I discussed earlier in the chapter. If a physical particle is a particle that has only physical-science properties, then naïve physicalism as defined here is an extreme form of outwardism. There is no reason to think that physical particles in this sense exist at all, let alone that they are all that exists. On the other hand, if a physical particle is a particle that has some physical-science properties, then naïve physicalism as defined here might be compatible with the quadrants view. My main reservation in saying this is that I don't think physical particles retain the form of particles when they fuse or merge together to make electronuclear fabric. However, Searle's wording "physical particles with their properties and relations" might be flexible enough to accommodate this concern

Another example is Galen Strawson, who describes his understanding of "physicalism" this way:

> I take physicalism to be the view that every real, concrete phenomenon in the universe is ... physical. It is a view about the actual universe, and I am going to assume that it is true. For the purposes of this paper I will equate 'concrete' with 'spatio-temporally (or at least temporally) located', and I will use 'phenomenon' as a completely general word for any sort of existent. (Strawson 2006, 3; his three dots)

This is not a complete definition because it leaves the key word "physical" unexplained. That explanation comes a few pages later:

> How can I say that 'physicalism' is an acceptable name for my position? Because I take 'physical' to be a natural-kind

term whose reference I can sufficiently indicate by drawing attention to tables and chairs and—as a realistic physicalist—experiential phenomena. The physical is whatever general kind of thing we are considering when we consider things like tables and chairs and experiential phenomena. It includes everything that concretely exists in the universe. (Strawson 2006, 8)

Strawson's physicalism is clearly different from physical-science physicalism, which he disparagingly dubs "physicSalism." It is also tremendously vague, which is one reason why it has no obvious incompatibility with the quadrants view. In fact, I am hard pressed to see how Strawson's physicalism differs from the tautology that everything real and concrete is concrete and real. The quadrants view is compatible with all tautologies.

These ambiguities of "physicalism" line up roughly with the ambiguities of "materialism" that I discussed in chapter 8. Both of these terms can be understood in a way that is compatible with the quadrants view, but for most writers who use them they denote extreme forms of outwardism that are sharply at odds with the quadrants view. Terms that are ambiguous in such hugely consequential ways are best avoided.

10

Generating Experience

In chapter 7, I gave several reasons to take seriously the possibility that electronuclear objects have remote properties. I also introduced the two-quadrant foundation hypothesis, which says that experiential properties are joint products of outwardly observable properties and remote properties. This chapter describes the two-quadrant foundation hypothesis in detail.

One can draw a distinction between the fundamental properties of the smallest electronuclear objects and the evolved properties of living electronuclear systems. Crossing this distinction with the distinction between outwardly observable properties and remote properties yields the four categories of nonexperiential properties in the following table:

	Outwardly Observable	Remote
Evolved	Examples: details of brain anatomy, electric pulse propagation in nerve cells.	No examples are available.
Fundamental	Examples: mass, electric charge.	No examples are available.

This table is an expansion of the bottom, nonexperiential row of the table of four quadrants that I presented in chapter 7. To make clear the relation between the two tables, I have shaded the outward quadrant in both of them. For reasons that I will now explain, I think it is plausible that properties in

110

all four of these nonexperiential categories are involved in the generation of experiential properties.

Evolved outwardly observable properties are involved. There is very strong evidence that certain evolved outwardly observable properties of the nervous system play a role in generating experience—properties such as the propagation of electrical signals in nerve cells and the diffusion of neurotransmitter chemicals across synapses. This evidence includes correlations of two kinds. First, as far as we can tell the only systems that have experiential properties are living animals that have nervous systems with these sorts of outwardly observable properties. Second, in human beings extensive correlations have been demonstrated between specific inwardly observable experiential properties and specific outwardly observable properties of the nervous system such as those revealed through electrical recordings, biochemical tests, and brain-imaging studies. If none of these outwardly observable properties of the nervous system are involved in the generation of particular features of experience, then all these correlations are bizarre coincidences.

Fundamental outwardly observable properties are involved. Evolved outwardly observable properties of the nervous system depend, at least in part, on certain fundamental outwardly observable properties of electrons and molecular nuclei. This dependence has at least two aspects. First, these fundamental outwardly observable properties of electrons and molecular nuclei play a key role in forming the chemical bonds that hold together larger electronuclear systems such as ions, neurotransmitters, neurons, sense organs and brains. Second, these fundamental outwardly observable properties of electrons and molecular nuclei play a key role in the transmission of electrical signals within neurons and chemical signals between neurons, which are such pervasive features of the brain's outwardly observable complexity. If experiential properties depend on certain evolved outwardly observable complexities of the brain, and if these evolved outwardly observable complexities of the brain depend in turn on certain fundamental outwardly observable properties of electrons and molecular nuclei, then, by the transitivity of dependence, experiential properties depend indirectly on these fundamental outwardly observable properties of electrons and molecular nuclei.

It is also possible that experiential properties have a direct dependence on certain fundamental outwardly observable properties of electrons and molecular nuclei. Since all the electrons and molecular nuclei throughout the brain have these fundamental outwardly observable properties, these properties are present throughout the regions of the brain where experience is generated—wherever these regions may be.

Evolved remote properties are involved. In chapter 7 I suggested that certain evolved remote properties might play a role in the storage format of sensory memories. Likewise, the same or other evolved remote properties might play a role in generating experiential properties. Any evolved remote properties that a living electronuclear system has are candidates for participation in the various things that take place in that system.

Fundamental remote properties are involved. Evolved remote properties that play a role in the generation of experience would presumably depend in part on fundamental remote properties of electrons and molecular nuclei. This would give experiential properties an indirect dependence on fundamental remote properties of electrons and molecular nuclei, by the transitivity of dependence.

It is also possible that experiential properties have a direct dependence on certain fundamental remote properties of electrons and molecular nuclei. If electrons and molecular nuclei have fundamental remote properties, then these fundamental remote properties are present throughout the brain, just as fundamental outwardly observable properties are. They are present in cell bodies, dendrites, axons, synapses, intercellular fluids, and so on. Thus, an experience-generating collaboration of properties that includes fundamental remote properties could occur anywhere in the brain.

How plausible is it that fundamental and/or evolved remote properties play a role in the generation of experiential properties? In chapter 7 I gave several reasons for believing that electronuclear objects have remote properties. Taking that to be the case, here are three reasons for believing that some of the remote properties of living electronuclear systems play a role in the generation of experiential properties.

First, if remote properties exist, why would they *not* play a role in generating experiential properties? I cannot think of any reason why all the work of generating experiential properties should be reserved for properties that happen to be outwardly observable. That would be a bizarre coincidence, if remote properties are also present. Playing a role in generating experiential properties would seem to be equally compatible with being outwardly observable and with not being outwardly observable.

Second, experiential properties and remote properties have something significant in common—not being outwardly observable. This suggests that experiential properties might inherit this characteristic from the remote properties that play a role in generating them. If this is the case, the generation of experience has a kind of continuity that it does not have if only outwardly observable properties are involved. The dawning of experience in the course of the evolution of life on earth would be a less abrupt, less radical development.

Third, if experience has remote causes that are beyond the reach of outward observation in just the way that experience itself is, and also beyond the reach of inward observation, then there is a straightforward explanation for the sense of futility that many people feel in trying to understand how experience is generated. People feel this sense of futility because these attempts are in fact futile. They are futile because some of the causes of experience are properties to which we can have no cognitive access. There is no way to understand or explain how experience is generated if some of the contributing factors are beyond the reach of cognition.

The two-quadrant foundation hypothesis is consistent with the possibility that some experiential properties are joint products of outwardly observable properties, remote properties, and other experiential properties. A causal configuration such as this might occur in cases where inward observation gives the impression that one experiential property leads to another. For example, seeing or hearing something sometimes triggers a thought or an emotional reaction. In many cases, however, experiential properties occur that do not have other experiential properties involved in their immediate causation. This is true of most sensory experience. According to the two-quadrant foundation hypothesis, for example, the visual experience that springs into existence each time you open your eyes is jointly produced by outwardly observable properties and remote properties of various parts of your body, without the assistance of any immediately preceding experiential properties. This sort of case, in which an experiential property is generated entirely by a combination of nonexperiential properties, is the focus of this chapter.

The two-quadrant foundation hypothesis is a joint causation hypothesis. But the joint causation described by this hypothesis differs from familiar examples of joint causation in the following important respect. Ordinarily, if joint causation is at work one can demonstrate the roles of the respective causal factors by removing one factor while leaving the others in place. By running your car until the gas tank is empty, you can demonstrate that gasoline plays a crucial role in making your car go. By placing a burning match in a vacuum, you can demonstrate that air plays a crucial role in sustaining a flame. By keeping the males away from the females, you can demonstrate that males play a crucial role in the reproductive process. However, one cannot demonstrate the experience-generating collaboration of outwardly observable properties and remote properties in this way because all the participating properties belong to the indissoluble property package of a single electronuclear object.

An electronuclear object has a plenary nature that can be thought of as a package of properties. All objects of a given type have similar property packages. For example, every electron has a mass, a charge, a spin, and more. There is no way to shave a single property off the plenary nature of an electron to create an object that is like an electron in every way except that it lacks that one property. The same holds for molecular nuclei, molecules, brains, and electronuclear objects in general. As objects become larger and more complex, opportunities exist to modify or destroy them, thereby creating different objects that have different property packages. But one cannot shave off single properties, leaving the rest of an object's property package intact. There is no such thing as property surgery. The property packages of electronuclear objects cannot be broken up.

This principle applies without regard to quadrants. If electrons have remote properties, then their remote properties are part of the indissoluble electron property package. There is no experimental manipulation that can address the question "What would happen if all the outwardly observable properties of electrons were retained while all the remote properties were removed?" Wherever there is an electron, wherever human observers detect an electron by means of its outwardly observable properties, there is an indissoluble electron property package, remote properties included. The same holds for molecular nuclei, molecules, brains, and electronuclear objects in general.

Another way to make the same point is to imagine what an instruction booklet for making an animal that has experience could say. It could not say "Take these outwardly observable properties and these remote properties, combine and mix well." It could not say this because we cannot get our hands on the outwardly observable properties in isolation from the remote properties. The outwardly observable properties and the remote properties always exist together, being inseparable aspects of indissoluble property packages. So the instruction booklet would have to say something like "Combine hydrogen, oxygen, carbon, and assorted other chemical elements as explained below." In following these instructions, one would at every moment be handling the indissoluble property packages that constitute the plenary natures of the electronuclear objects that the instructions mention.

This completes my basic account of the two-quadrant foundation hypothesis. Next I add perspective and detail under two headings—the explanatory gap and supervenience.

<>

The phrase "the explanatory gap," first used in this context by Joseph Levine in 1983, combines an evident fact with a gratuitous assumption. The evident fact is that there is something we would like to explain but at present cannot: we cannot at present explain how experience in general, or any particular feature of experience, is generated from nonexperiential ingredients. We do not understand how nature's technology for generating experience works. The gratuitous assumption is reflected in the word "gap." This word suggests a certain form of explanation, which consists of a gap-spanning "bridge" that connects outwardly observable, physical-science properties of the brain to inwardly observable features of experience. Here are passages written by two of the many writers who aspire to "bridge the explanatory gap":

> So we will attempt to bridge the explanatory gap by working from both ends—neural and experiential. (Taylor 1999, 53; note his outwardly skewed usage of "neural.")

> Eventually, a theory that bridges the explanatory gap, that explains why activity in a subset of neurons is the basis of (or, perhaps, is identical to) some particular feeling, is required. (Koch 2004, 18)

Since it makes no sense to attribute a gap-spanning form to something that does not exist, the phrase "explanatory gap" suggests that an explanation exists. This suggestion is gratuitous, and according to the quadrants view it is false. According to the quadrants view, it is impossible to produce a correct explanation of how any experiential property is generated, let alone a correct explanation that has the form of a gap-spanning bridge, because the experience-generating process involves remote properties that can never be recognized by an explainer and used in an explanation.

Consider the following analogy. On lazy summer days, children sometimes entertain themselves by using a magnifying glass to set dry leaves or scraps of paper on fire. Imagine a naïve analyst who does not notice that sunlight plays a role in this pastime. He racks his brain trying to understand how a flame could spring from a thick circular piece of glass; the cause and the effect seem disconnected and so dissimilar. He declares that there is an explanatory gap between the magnifying glass and the flame. It is easy to appreciate that this analyst is misguided. The correct explanation involves another causal factor—solar radiation—which works with the magnifying glass to produce the flame. The explanation does not take the form of a gap-spanning bridge from the magnifying glass to the flame. If the two-quadrant foundation hypothesis is correct, then those who speak of "the explanatory gap" are misguided in much the same way. They mistakenly assume that inwardly observable features of experience spring from outwardly observable, physical-science properties of the nervous system in a way that calls for explanation by a gap-spanning bridge. Actually, other causal factors are involved—remote properties that work with outwardly observable properties to produce experiential properties. When people puzzle over the correlations between inwardly observable features of experience and outwardly observable properties of the brain, what they are missing is not a gap-spanning bridge between the two, but rather another set of causal factors that work with the outwardly observable causal factors to generate features of experience.

Note that this analogy has two significant limitations. First, the production of the flame is an ordinary case of joint causation in which it is possible to demonstrate the respective roles of the two causal factors by removing one of them. A sunny day without a magnifying glass or a magnifying glass without sunlight produces no flame; only a magnifying glass with sunlight streaming through it produces a flame. In contrast, the outwardly observable properties and the remote properties that jointly produce experiential properties belong to a single indissoluble property package, so it is not possible to separate them experimentally. Second, in the case of the flame the factor that the naïve analyst overlooks is available for study, making an explanation of how the flame is produced possible. But with the generation of experiential properties, the remote causal factors that users of the phrase "the explanatory gap" overlook are not available for study, so no explanation of how experiential properties are generated is possible. What is possible is an explanation of why the desired explanation is not possible. The two-quadrant foundation hypothesis provides that.

In a discussion of the so-called explanatory gap, Jaegwon Kim writes:

> When we are faced with an explanatory demand "Why
> p?", there are two ways of meeting it: first, we can provide
> a correct answer to this why-question; second, we can show

that *p* is false and that there is nothing to be explained. (Kim 2010, 276)

This list of possible responses to a question of the form "Why *p*?" is crucially incomplete. A third possible response is to acknowledge that there is good reason to accept *p* and then explain why it is that we will never be able to explain why *p* is true. According to the quadrants view, this is the correct response whenever *p* is a well-confirmed statement of correlations between inwardly observable experiential properties and outwardly observable properties of the brain. Someone might object that a response of this form does not successfully "meet the demand" for an explanation. But this would be to impose an arbitrary restriction on what counts as meeting a demand. It is logically possible that human beings are able to see that certain things are true while being blocked from ever being able to explain why they are true. Our cognitive powers might be limited in that way. Accordingly, there is no reasonable basis for a blanket rejection of all responses of this form. If the explanation of why there can be no explanation of *p* is cogent and compelling, the explanatory demand has been "met."

Levine's phrase "the explanatory gap" is not wholly responsible for sending people down the gap-bridging path. There were people looking for an explanation of this form before Levine introduced this phrase, and in all likelihood there would be people looking for an explanation of this form today even if this phrase had not been introduced. The phrase "the explanatory gap" merely encapsulates and encourages this way of thinking; it is not the source of it.

<>

The word "supervenience" is used in different ways by different writers. My focus here is the work of Jaegwon Kim, in which the word "supervenience" plays a central role.

Kim says that he believes in "mind-body supervenience." This phrase suggests to me a body-plus view according to which a human being consists of a body plus something else, which supervenes on the body. The following explanation of this phrase also suggests a body-plus view:

> For present purposes we will not need an elaborate statement
> of exactly what mind-body supervenience amounts to. It will

117

> suffice to understand it as the claim that what happens in our
> mental life is wholly dependent on, and determined by, what
> happens with our bodily processes. (Kim 2005, 14)

Understood in this way, I reject mind-body supervenience for the reasons given in chapter 2. According to the quadrants view, what happens in a person's mental life is an aspect of certain electronuclear, bodily processes; it is not something additional that depends on bodily processes. However, it is not clear that this is a correct understanding of Kim's view. As I noted in chapter 8, Kim mixes body-plus and body-only descriptions in a way that makes it difficult to decide where he stands on this issue. It is possible to read him as advocating a body-only view according to which the body's "mental properties" supervene on the body's "physical properties." This would make the phrase "mind-body supervenience" misleading, but it is a possible way to read Kim, all things considered. It is also the only way to understand Kim's supervenience theory that requires further discussion, so this is how I will understand it here.

Kim makes two related supervenience claims that tend to get blended together in his writing. The first is that the relation between "mental properties" and "physical properties" falls under a broad notion of supervenience that "merely states a pattern of property covariation between the mental and the physical and points to the existence of a dependency relation between the two." He elaborates this as follows:

> … supervenience is not a *type* of dependence relation—it is
> not a relation that can be placed alongside causal dependence,
> reductive dependence, mereological dependence, dependence
> grounded in definability or entailment, and the like.
> Rather, any of these dependence relations can generate the
> required covariation of properties and thereby qualify as a
> supervenience relation. (Kim 1998, 14; his emphasis)

In this passage, Kim seems to use the word "supervenience" as a synonym or near-synonym of "dependence." He uses both words, but it does not seem that he needs both of them. The message of the passage is only that "mental properties" are *somehow or other* dependent on "physical properties." The second, much more specific claim Kim makes is that there is *a particular type* of dependence relation between "mental properties" and "physical properties."

This particular dependence relation connects certain properties in pairs and is necessary, instantaneous, and noncausal:

> Mental properties *supervene* on physical properties, in that necessarily, for any mental property M, if anything has M at t, there exists a physical base (or subvenient) property P such that it has P at t, and necessarily anything that has P at a time has M at that time. (Kim 1998, 9; his emphasis)

> ... in general, the relation between base properties and supervenient properties is not happily construed as causal. For one thing, the instantiations of the related properties are wholly simultaneous, whereas causes are standardly thought to precede their effects ... (Kim 1998, 44)

Kim draws diagrams in which this particular mental-physical dependence relation has a pictorial representation. The diagrams have a horizontal time axis. In them, causation within a person is represented by horizontal arrows pointing from left to right parallel to the time axis, while mental-physical supervenience within a person is represented by vertical arrows pointing upwards from a "physical property" designated P to a "mental property" designated M.

Since Kim's category of "mental properties" includes experiential properties, each of these claims has a corollary that concerns the dependence of experiential properties on "physical properties." The corollary of his general claim is that experiential properties depend somehow or other on "physical properties." The corollary of his specific claim is that experiential properties depend in a necessary, instantaneous, and noncausal way on "physical properties." These corollaries can be compared with the quadrants view.

Kim's view differs from the quadrants view in four noteworthy respects, each of which raises a problem for Kim's view. First I discuss three problems that pertain only to his more specific supervenience claim. The fourth problem pertains to both of his supervenience claims.

One problem with Kim's specific supervenience claim is the idea that each "mental property" depends directly on only one "physical property," which is "presumably, a complex neural property" (1998, 9). That one "physical property" can in turn have dependencies on any number of other "physical properties," but it is a distinctive feature of Kim's view that for each instance of a "mental property" there is one and only one "physical property" that is a kind of pedestal for it. I do not believe that Kim gives any reason why supervenience must have this structure; it seems to be an arbitrary stipulation. In any case, the quadrants view includes no such structural constraint. According to the quadrants view, experiential properties have causal dependencies on many nonexperiential properties, and the way in which the causally contributing properties work together to produce experiential properties is left open. For example, as explained at the beginning of this chapter, a given experiential property could have direct causal dependencies on evolved properties of an animal's nervous system and also on fundamental properties of the electrons, protons, and neutrons of which that nervous system is made.

A second problem with Kim's specific supervenience claim is that the nature of this particular type of supervenience relation is obscure. Kim describes this type of supervenience using ostensibly noncausal terms such as "guarantee" and "fix," as in the following passages:

> ... every mental property has a physical base property that *guarantees* its instantiation. (Kim 1998, 10; my emphasis)

> ... the mental nature of a thing *is entirely fixed by* its physical nature. (Kim 1998, 11; my emphasis)

These descriptions raise an important question that Kim does not answer. *How* does a "physical property" guarantee and fix an experiential property, if not through causation? *By what noncausal means* does a "physical property" guarantee and fix an experiential property? I find it impossible to conceive of a noncausal way in which a "physical property" can guarantee and fix an experiential property. I can't say that this requirement is flatly self-contradictory, but it comes pretty close. The obscurity of noncausal supervenience is absent from the quadrants view, because according to the quadrants view experiential properties are *causally* dependent on nonexperiential properties.

A third problem is that there is no evidence—worse, there can never be any evidence—for the existence of instantaneous noncausal supervenience relations of the kind that Kim describes. The reason is that there is no way to distinguish such a relation from very fast causation, which is well known to be commonplace in electronuclear fabric. The time that it takes for electrons and molecular nuclei to undergo the rearrangement that we call a chemical reaction is several orders of magnitude less than the time that it takes for a person to notice and report the onset of a given feature of experience. In order to make a case that this feature of experience has an instantaneous noncausal dependence on a "physical property," one needs some reason to favor that hypothesis over the less exotic alternative that the experiential property is generated causally with a speed comparable to that of a chemical reaction. The relative crudeness of our ability to time inwardly observed changes makes it impossible to produce any evidence that favors the more exotic hypothesis of instantaneous noncausal supervenience.

The obscurity of the idea of instantaneous noncausal supervenience combined with the impossibility of producing any evidence for the reality of such a relation adds up to a strong case that there is no such relation in us. On the other hand, there is extensive evidence that electronuclear fabric is home to very fast causation of various types, so the generation of experiential properties by very fast causation fits into a known pattern.

I have a conjecture as to why Kim finds the idea of instantaneous noncausal supervenience attractive. My conjecture is that this idea is rooted in an illusion that is produced whenever very fast causation is combined with persisting causal conditions. Consider the example of an ordinary light bulb. Electric current flows continuously through the filament of the bulb, causing the filament to give off a continuous flux of photons—light. The radiating photon flux depends on the filament and the electric current by way of very fast causation; there is no relation of instantaneous noncausal supervenience here. However, if the bulb stays on for a while, during most of that time the conditions in the filament of the bulb are precisely those that *can* produce the photon flux that exists at that very instant. This might lead one to think that there is an instantaneous and ongoing noncausal supervenience relation between the photon flux and the electric current. In fact, however, the photon flux right now was caused by the conditions in the filament a millionth of a second ago, while the conditions in the filament right now are in the process of producing the photon flux that will be radiating from the bulb a millionth of a second from now.

Another similar example, a bit less close to home, is the gravitational attraction between the earth and the moon. It is natural to think of this attraction as depending instantaneously on the masses of the earth and the moon, and thus to be supervenient on those masses in Kim's specific sense. This is how Isaac Newton thought of it. According to today's physics, however, all gravitational influences travel at the speed of light, from which it follows that the gravitational attraction between the earth and moon right now is causally dependent on their masses approximately 1.3 seconds ago. Since their masses have not changed much in the last 1.3 seconds, an illusion of instantaneous dependence is produced.

According to the quadrants view, a living animal's experiential properties are generated continuously by very fast causation, with a speed comparable to that of a chemical reaction or the generation of a photon flux by the electric current in the filament of a light bulb. While I am awake, the causal conditions of wakefulness continuously generate my wakeful state. While my nose is itching, the causal conditions of that itchy feeling continuously generate that itchy feeling. And so on for all the features of my experience. Crudely timed correlation studies connecting inwardly observed experiential properties with outwardly observed properties of the brain can foster the illusion of instantaneous noncausal supervenience, but in fact the dependence is a case of ongoing very fast causation.

The fourth difference between Kim's view and the quadrants view is that Kim's supervenience claims give no role to remote properties. All the dependencies of "mental properties" are on "physical properties," which Kim evidently understands as outwardly observable physical-science properties. Kim's conceptions of supervenience—both general and particular—thus have the same problem as the idea that there is an explanatory gap between physical-science properties and mental properties. If the quadrants view is correct, all of these ideas involve the same incomplete and outwardly skewed picture of the dependencies of experiential properties. According to the two-quadrant foundation hypothesis of the quadrants view, experiential properties are jointly produced by outwardly observable physical-science properties and remote properties, all of which belong to the same indissoluble property packages of the body's electronuclear fabric. If it seems that experiential properties depend only on physical-science properties or that the existence of experiential properties should be explainable in terms of physical-science properties, this is an illusion that is rooted in the fact that there are causally contributory remote properties that always accompany the causally contributory physical-science properties but are not cognitively accessible.

Given a broad notion of supervenience that subsumes very fast causation, one could say that according to the quadrants view experiential properties supervene on a combination of outwardly observable properties and remote properties. Or, to use another one of Kim's idioms, one could say that according to the quadrants view experiential properties have a two-quadrant "supervenience base" that includes outwardly observable properties and remote properties. However, I prefer not to say these things, for two reasons. First, they add nothing of substance to what I have already said. And second, they have the downside of inviting confusion with Kim's specific idea of instantaneous and noncausal supervenience, which is wholly erroneous according to the quadrants view. Being different from Kim's supervenience view in all the ways that I have spelled out here, the quadrants view is best described without using the word "supervenience."

As an icon for the causal configuration of the two-quadrant foundation hypothesis, I propose the Greek letter π. The horizontal stroke represents experiential properties, whether inwardly observable or not, and the two vertical strokes represent the causally contributory physical-science properties and remote properties, respectively. The separateness of the two vertical strokes must not be taken to suggest any separation between the physical-science properties and the remote properties, however. An even better icon would be a Greek letter π with its two legs wrapped around each other, indicating that all the causally contributory properties belong to the indissoluble property packages of the body's electronuclear fabric.

11

The Quadrants and Science

Many writers expect the advance of science to lead to a scientific explanation of how experience is generated, and some have tried to construct such an explanation based on current scientific knowledge. According to the quadrants view, there can be no such explanation because scientists cannot discover the remote properties that play a role in generating experience. Scientists cannot study, let alone understand, the way remote properties work together with outwardly observable properties to produce experiential properties.

One can certainly study experience scientifically. Scientists have discovered many interesting relationships between various features of experience and various outwardly observable properties, including outwardly observable properties of the nervous system, the sense organs, sensory stimuli, behavior, ingested or injected substances, and more. As long as scientists continue to pursue research in this area, they will continue to make interesting and useful discoveries of this sort. Over time, it is even possible that scientists will discover every outwardly observable property that plays a role in generating each and every inwardly observable feature of human experience. But what is not possible, according to the quadrants view, is that someone will parlay such discoveries into a correct explanation of how these features of experience are generated. This is not possible because nature's technology for generating features of experience involves remote properties, which are beyond the reach of scientific study.

This chapter examines a variety of claims that conflict with this implication of the quadrants view. The writers considered are Gerald Edelman, John G. Taylor, Christof Koch, John Searle, Stanislas Dehaene, Herbert Feigl, Jeffrey Gray, and Francis Crick.

<>

In a series of books, Gerald Edelman presents what he calls a "biological theory of consciousness." Recall my discussion in chapter 8 of the ambiguity of the word "biological." Edelman uses the word "biological" in the outwardly skewed sense that limits his explanatory resources to the outward-observation-based *science of biology*. According to the quadrants view, it is not possible to construct a correct explanation of how experience is generated in this way because the causal contribution of remote properties is left out of account. The following critique of Edelman's theory is consistent with this claim.

A central feature of Edelman's theory is his distinction between "primary consciousness" and "higher-order consciousness," which he describes in passages such as the following:

> I have made a distinction, which I believe is a fundamental one, between primary consciousness and higher-order consciousness. Primary consciousness is the state of being mentally aware of things in the world—of having mental images in the present. But it is not accompanied by any sense of a person with a past and future. It is what one may presume to be possessed by some nonlinguistic and nonsemantic animals ... In contrast, higher-order consciousness involves the recognition by a thinking subject of his or her own acts or affections. It embodies a model of the personal, and of the past and future as well as the present. It exhibits direct awareness—the noninferential or immediate awareness of mental episodes without the involvement of sense organs or receptors. It is what we as humans have in addition to primary consciousness. We are conscious of being conscious. (Edelman 1992, 112)

> Primary consciousness—the ability to generate a mental scene in which a large amount of diverse information is integrated for the purpose of directing present or immediate behavior—occurs in animals with brain structures similar to ours. Such animals appear able to construct a mental scene but, unlike us, have limited semantic or symbolic capabilities and no true language. Higher-order consciousness is built on the foundations provided by primary consciousness and is accompanied by a sense of self and the ability in the waking

state explicitly to construct and connect past and future scenes. In its most developed form, it requires a semantic capability and a linguistic capability. By necessity, only individuals who are endowed with higher-order consciousness can report conscious states and speak about consciousness; they are conscious of being conscious. (Edelman and Tononi 2000, 103–104)

According to Edelman, primary consciousness is an evolutionary precursor of higher-order consciousness, and a foundation for it.

The part of Edelman's theory that is of interest here is his account of the origin and foundations of primary consciousness. This account involves three main factors. Before the evolutionary dawn of primary consciousness, there were animals that had 1) sensory processes that involve "perceptual categorization in real time" and 2) a kind of memory that he calls "value-category memory." Primary consciousness came into being with the addition of 3) two-way "reentrant" neural interaction that connects these sensory processes and this kind of memory:

A third and critical evolutionary development provides a sufficient means for the appearance of primary consciousness. This is a special reentrant circuit that emerged during evolution as a new component of neuroanatomy. This circuit allows for continual reentrant signaling between the value-category memory and the ongoing global mappings that are concerned with perceptual categorization in real time. An animal without these new reentrant connections can carry out perceptual categorization in various sensory modalities and can even develop a conceptual value-category memory. Such an animal cannot, however, link perceptual events into an ongoing scene. With the appearance of the new reentrant circuits in each modality, *a conceptual categorization of concurrent perceptions* can occur before these perceptual signals contribute lastingly to that memory. This interaction between a special kind of memory and perceptual categorization gives rise to primary consciousness. (Edelman 1992, 119; his emphasis)

In brief, an animal's primary consciousness at a given moment is the product of ongoing interaction between sensory processes that are triggered by current sensory stimuli and a certain kind of long-term memory that was jointly shaped by past sensory processes and "value."

There is some instability in Edelman's delineation of the class of animals that have primary consciousness. In *Bright Air, Brilliant Fire* he bounds this class in the following way:

> If the brain systems required by the present model represent the *only* evolutionary path to primary consciousness, we can be fairly sure that animals without a cortex or its equivalent lack it. An amusing speculation is that cold-blooded animals with primitive cortices would face severe restrictions on primary consciousness because their value systems and value-category memory lack a stable enough biochemical milieu in which to make appropriate linkages to a system that could sustain such consciousness. So snakes are in (dubiously, depending on the temperature), but lobsters are out. If further study bears out this surmise, consciousness is about 300 million years old. (Edelman 1992, 123; his emphasis)

But in books written both earlier and later, he puts the evolutionary dawn of primary consciousness perhaps only half that long ago, thereby excluding snakes along with all other reptiles:

> Consciousness of the first kind, or primary consciousness, arose in such vertebrate species (probably with birds, but almost certainly with mammals, as a function of newly evolved reentrant brain structures. (Edelman 1989, 263)

> At a point in evolutionary time corresponding to the transitions between reptiles and birds and reptiles and mammals, a critical new anatomical connectivity appeared. Massively reentrant connectivity arose between the multimodal cortical areas carrying out perceptual categorization and the areas responsible for value-category memory. This evolutionarily derived reentrant connectivity is implemented by several grand systems of corticocortical fibers linking one part of

the cortex to the rest and by a large number of reciprocal connections between the cortex and the thalamus.... All these thalamocortical structures and their reciprocal connections acting together via reentry lead to the creation of a conscious scene. (Edelman and Tononi 2000, 107–108)

We do not know at what point in evolutionary history primary consciousness first arose. However, by comparing homologous neural structures required for its expression in humans and other vertebrates (for example, a thalamocortical system and ascending value systems along with certain behavioral patterns), we can put forth a tenable conjecture that primary consciousness appeared in vertebrates first at the transition between reptiles and birds, and second at the transition between reptiles and mammals. (Edelman 2004, 132)

This instability in Edelman's placement of the dividing line between animals with and animals without primary consciousness is of no importance to my critique. What is very important, however, is that on either placement of the dividing line there is a very large class of animals that do not have even primary consciousness. Insects, crustaceans, and fish are definitely in this class, according to the theory, and perhaps also reptiles and amphibians. Edelman's theory thus divides animals into three classes. Human beings belong to a relatively small class of animals that have both primary consciousness and higher-order consciousness. There is a much larger class of animals that have primary consciousness but not higher-order consciousness. And there is another large class of animals that have neither primary consciousness nor higher-order consciousness.

Edelman says very little about animals that do not have primary consciousness. This class of animals is always in the background as he swings his focus back and forth between animals with only primary consciousness and animals with both primary and higher-order consciousness. As a result, the following question of interpretation arises. Is Edelman's view that the evolutionary dawn of primary consciousness is the evolutionary dawn of experience, or is his view that experience dawned earlier and the dawn of primary consciousness is a stage in the development of experience? In other words, does Edelman's theory say that all animals without primary consciousness are experientially blank, or does it say that some of these animals have experience of a simpler sort that does not amount to primary consciousness as he defines it? Edelman's treatment of animals that do not have primary consciousness is so sketchy that it is very difficult, if not impossible, to answer this crucial question. His writing does include relevant indications, but they point in both directions and therefore serve only to underscore the question.

Here are four indications that Edelman thinks of the evolutionary dawn of primary consciousness as the evolutionary dawn of experience.

First, this seems to be the suggestion of the term "primary consciousness." If experience dawned earlier, there would be nothing especially primary about primary consciousness, so it would make more sense to give it a purely descriptive name such as "scene recognition consciousness."

Second, Edelman associates the phrase "primary consciousness" with a variety of other phrases, including "conscious experience," "perceptual experience," "conscious sensation," "phenomenal experience," and "qualia," all of which suggest to me that he thinks of himself as discussing experience in general.

Third, he never explicitly attributes experience to any animal that lacks primary consciousness.

Fourth and last, he seems to regard his theory as having a kind of importance that it would not have if it were merely a theory about certain stages in the evolution of experience, whose prior existence is assumed. Granted, a theory about certain stages in the evolution of experience could be interesting and significant, but I do not think it would merit the fanfare that is implicit in Edelman's writing.

On the other hand, here are three indications that Edelman does not think of the evolutionary dawn of primary consciousness as the evolutionary dawn of experience.

First, his theory gives a key role to an evolutionary precursor of primary consciousness that he calls "value-category memory." For me, the thought that an animal *values* something is difficult to separate from the thought that it has experience of some sort. This linkage arouses the suspicion that Edelman is assuming the existence of certain kinds of value-related experience in animals that do not have primary consciousness. There is support for this suspicion in descriptions of value-category memory such as the following:

> Learning must be related to evolved species-specific hedonic, consummatory, appetitive, and pain-avoiding behaviors that reflect ethologically determined values. (Edelman 1989, 93)

> The activities of hedonic centers and various responses to nociceptive signals provide the basis for the values that, together with memory, result either in unchanged behavior or in an alteration of behavior that leads to learning. (Edelman 1989, 112)

> These are the ascending systems, which my colleagues and I have called value systems because their activity is related to rewards and responses necessary for survival. (Edelman 2004, 25)

The words "hedonic," "appetitive," "pain," "nociceptive," and "reward" are commonly understood to concern aspects of experience. In particular, these words suggest features of experience such as feelings of hunger and thirst, warmth and cold, burning and itching, sweetness and juiciness. If Edelman is not assuming the existence of such features of experience in animals that lack primary consciousness, then he needs to explain how to understand these words nonexperientially. I think it is highly questionable whether Edelman can explain "value-category memory" in a way that does not presuppose a considerable variety of value-related experience in animals that lack primary consciousness.

Second, Edelman never explicitly says that animals without primary consciousness have no experience. In describing animals that lack primary consciousness, he uses abstract phrases that are compatible with the absence of experience, such as "perceptual categorization" and "sampling of the environment," but he never comes right out and says that insects, crustaceans, fish, and perhaps reptiles and amphibians, are as experientially blank as a fence post.

Third, the view that all these animals without primary consciousness are experientially blank is a controversial one that is sure to arouse strong objections, yet Edelman gives no hint that he thinks of himself as advancing a controversial view. Owen Flanagan writes, "Lobsters, I suspect, are dim experiencers" (1991, 325). I suspect that many people share this suspicion. If Edelman does not, one might expect him to acknowledge its appeal and to say something by way of rebuttal. In the following passage, Edelman gives a passing mention to bees:

> It is true that learning and communication systems arose in evolution well before primary consciousness. Organisms such

> as bees or wasps can, for example, show remarkable adaptive
> behavior in groups that depends to some degree on individual
> variation. (Edelman 2004, 132)

If he thinks that bees build their hives, fly from flower to flower while never colliding with trees, and communicate by means of their famous waggle dance, all without the benefit of visual experience, one might expect him to acknowledge that some readers will find that difficult to believe and to say something to try to win over these doubters. Why does Edelman never address this predictable objection to the view that all these animals are experientially blank? One possible explanation is that he does not think that these animals are experientially blank, but only that their experience is of a simpler sort that does not count as primary consciousness by his definition.

I see no way to settle this question. Accordingly, I will consider Edelman's theory under both interpretations.

Suppose first that the theory attributes experience to some animals that do not have primary consciousness. It thus casts the evolutionary dawn of primary consciousness as a stage in the evolution of already existing experience. Perhaps the theory says that the interaction of current sensory processes with long-term "value-category memory" upgrades the experience of simpler animals by adding certain "advanced" features such as the recognition of previously encountered objects, a sense of familiarity with one's surroundings, and a sense of where one is. By definition, it is the development of such "advanced" features that constitutes the dawn of primary consciousness. In this case, the theory does not even address the question of how experience is generated. Use of the term "primary consciousness" makes it seem that Edelman is discussing this question, but in fact he is not.

Now suppose the theory says that all animals without primary consciousness are experientially blank. It thus casts the evolutionary dawn of primary consciousness as the evolutionary dawn of experience. In this case, the theory addresses the question of how experience is generated, but it has the following three deficiencies.

First and most importantly, the theory gives no explanation of how the alleged foundations of experience give rise to experience. If the neural activity called "value-category memory" and the neural activity called "perceptual categorization in real time" are individually wholly nonexperiential, how does a reentrant interaction between them manage to be experiential? One can understand how an interaction between long-term memory and signals coming from sense organs could make the difference between an animal that recognizes previously encountered objects and one that does not, or the difference between an animal that has a sense of where it is and one that does not, but there is no apparent connection between such an interaction and the difference between an animal that has experience and an animal that is experientially blank. Moreover, without any explanation of how the interactions that Edelman discusses give rise to experience, there is no reason to believe that they do give rise to experience.

Second, the theory includes no rebuttal of the natural suspicion that there is some sort of sensory experience in any animal that has sense organs much like ours, behaves as if its sense organs tell it something about its surroundings, and is a product of the same evolutionary process that produced us. Why should I resist the temptation to believe that lobsters, bees, and countless other animals that evolved before mammals and birds have experience? I don't say that no rebuttal of this suspicion is possible. The point is that Edelman's failure to give such a rebuttal weakens the case for his theory.

A third deficiency, which will take a bit longer to discuss, is that any theory that makes experience dependent on long-term memory is going to have difficulty accounting for the onset of experience in the life cycle of an individual animal. In passages such as the following, Edelman describes primary consciousness that depends on an animal's memory of past *experience*:

> If an organism has primary consciousness and a concept system, however primitive, past *experiences* ... will lead to storage of changes that alter future behavior. (Edelman 1989, 257; my emphasis)

> Looked at from the inside, consciousness seems continually to change, yet at each moment it is all of a piece—what I have called "the remembered present"—reflecting the fact that *all my past experience* is engaged in forming my integrated awareness of this single moment. (Edelman 2004, 8; my emphasis)

> Imagine an animal with primary consciousness in the jungle. It hears a low growling noise, and at the same time the wind shifts and the light begins to wane. It quickly runs away, to a safer location. A physicist might not be able to detect any

necessary causal relation among these events. But to an animal with primary consciousness, just such a set of simultaneous events might have accompanied *a previous experience*, which included the appearance of a tiger. Consciousness allowed integration of the present scene with the animal's *past history of conscious experience*, and that integration has survival value whether a tiger is present or not. (Edelman 2004, 11; my emphasis)

An animal with primary consciousness can nonetheless discriminate and connect these objects and events through the memory of *its previous value-laden experience*. This ability enhances its survival value. (Edelman 2004, 56; my emphasis)

I can understand how the nature of the current experience of a grown animal might depend in part on its memories of its previous experience, but this cannot be the whole story because the claim that all experience owes its existence to memories of previous experience leads to an infinite regress. For Edelman to avoid this infinite regress, it must be the case that an animal's first experience is preceded by a type of long-term memory formation that is not memory *of experience*.

There is logical room to meet this requirement. Memory in the broadest sense combines a general resistance to change with a susceptibility to certain specific types of change; it need not be memory of experience. The most familiar example of memory that is not memory of experience is computer memory. Edelman himself gives several examples, including DNA that is open to mutations that it then retains (Edelman and Tononi 2000, 210), an adaptive immune system that permanently increases the numbers of specific types of antibodies in response to an infection (Edelman and Tononi 2000, 99–101), and the icy surface of a glacier that retains the system of channels that is carved into it by running water during thaws (Edelman 2004, 52–53). The trouble is that he never identifies a mode of memory *in animal brains* that is not memory of experience. There might be such a thing. If silicon chips, DNA molecules, immune systems, and glaciers can exhibit memory that is not memory of experience, then it may well be that brain tissue can too. But if it can, Edelman needs to explain how this experience-independent brain memory differs from the experience-dependent brain memory that we are all familiar with, and how it contributes to the start-up process for an individual animal's experience. He does neither.

For humans at least, Edelman seems to think that experience is present at birth:

> An animal, or even a newborn baby, with these dynamics and with primary consciousness will experience a scene but have no nameable self that is differentiable from within. (Edelman and Tononi 2000, 174)

From this it seems to follow that the start-up process for experience, with its dependence on experience-independent memory, is already underway in the womb. This deepens the mystery about how this start-up process is supposed to work. If Edelman intends primary consciousness to be coextensive with experience, I think it is doubtful that he can avoid the charge that there is an infinite regress at the heart of his theory.

In sum, on one interpretation Edelman's theory is not about the generation of experience at all, while on the other interpretation it has several major deficiencies. These deficiencies are 1) failure to provide an explanation of how experience is generated, 2) failure to defend against a serious objection, and 3) an apparent infinite regress in which all experience depends on an animal's memories of previous experience.

<>

The Race for Consciousness by John G. Taylor focuses on human consciousness; it does not share the evolutionary perspective of Edelman's books. But what Taylor says about human consciousness is similar to what Edelman says. Like Edelman, Taylor does not make it clear whether he thinks that "consciousness" is coextensive with experience or a narrower category. And like Edelman, Taylor maintains that consciousness depends on an interaction between current sensory input and long-term memory.

To make Taylor's theory relevant to the generation of experience, let's assume that he takes "consciousness" to be coextensive with experience. In this case, Taylor's theory, like Edelman's, has the deficiency that it gives no explanation of how the interaction of nonexperiential sensory input and nonexperiential long-term memory manages to produce experience; it simply states that this happens. Also like Edelman's theory, it seems to involve an infinite regress in which all experience owes its existence to memories of previous experience. This apparent circularity is illustrated by the following passages:

> What is clear so far is that consciousness involves memory structures or representations of the past of episodic, autobiographical, semantic, preprocessing, and emotional character. These structures are used to give conscious content to the input in a manner that endows that experience with meaning related to the past. Thus consciousness arises from the intermingling of recorded past experiences with incoming present activity ... (Taylor 1999, 37)

> In all, then, *the conscious content of an experience comprises the set of relations of that experience to stored memories of relevant past experiences*. Thus consciousness of the blue of the sky as seen now is determined by stored memories of one's experience of blue skies. (Taylor 1999, 125; his emphasis)

Unlike Edelman, Taylor considers objections to his theory, including this one that concerns the threat of an infinite regress:

> A reviewer of one of my papers on the model stated, in opposition to the theory, "A newborn baby, most people would agree, is a conscious being and yet he/she will initially be unable to relate to any 'somewhat similar past activity.'" This is an important objection to the emergent mind model, so I will answer it by carefully considering the facts. (Taylor 1999, 321)

Taylor responds to this objection, in part, as follows:

> ... the thesis of relational consciousness indicates that the level of conscious experience is determined by the level of past memories of the various sorts outlined so far. The level of consciousness will continuously increase from a very low level (provided by fetal experience) as the newborn begins to acquire such memories....

135

> The conclusion is that the claim that the newborn is conscious does not have much support. The level of consciousness that does exist is only very low and, it is suggested, arises from prenatal experience that is known to exist and be relevant to later responses. (Taylor 1999, 322–323)

I find this response inadequate in two respects. First, it seems to declare victory and admit defeat in the same breath. If the start-up process for experience begins with "fetal experience" or "prenatal experience" that is not based on memory, then memory is not at the root of the process, and the theory is wrong. Memory would serve instead as a kind of experience-enrichment device that needs an initial input of memory-independent experience to give it something to do. Second, I have difficulty accepting Taylor's claim that the consciousness of a newborn baby increases continuously from a "very low level." This is partly because the meaning of the phrase "level of consciousness" is not clear and partly because newborn babies seem to burst on the scene with experience in rather good supply. My exposure to newborn babies has been more in keeping with the famous supposition of William James that "the baby, assailed by eyes, ears, nose, skin, and entrails at once, feels it all as one great blooming, buzzing confusion" (2007, 488). A newborn baby's experience may well be chaotic and bewildering, but I have never had the impression that it is lacking in vividness or complexity. For both of these reasons, I do not think Taylor successfully rebuts the newborn baby objection. The reviewer who posed this objection saw very clearly the infinite regress that is implicit in Taylor's theory.

The newborn baby objection is a special case of a more general objection that concerns the lifelong theme of experiential novelty. Each feature of experience that occurs in a given life has a first occurrence in that life—a first taste of pineapple, a first view of a rainbow, a first orgasm, a first sense of grief. If such first occurrences depend on memories that were formed earlier in life, then that dependence is extremely obscure and needs to be explained. Taylor touches on the phenomenon of occasional novel experience in the following passage:

> If there is no past memory and the input is completely novel, special strategies are available to deal with it. One is that an animal will freeze, remaining stationary while watching the novel object carefully. It will then approach the object cautiously. In this way, it will gradually develop memory

structures that will be effective in dealing with what was initially a new and threatening object. (Taylor 1999, 37)

But here again, he seems to undercut his theory while claiming to support it. This passage does not sound like a description of an animal that is short on consciousness. On the contrary, it sounds like a description of an animal that is on high alert and attentive to numerous novel experiential details, despite—or maybe even because of—its lack of pertinent memories.

The thesis that experience owes its existence to long-term memory seems not only wrong, but an unlikely one to have well-educated advocates such as Edelman and Taylor. Clearly, long-term memory plays a role in shaping the character of a lot of experience. Memory enables us to recognize things, including objects, faces, voices, words, emotions, gestures, and more. Memory gives us a feeling of familiarity when we are in surroundings that we have been in before. Memory enables us to recollect and reminisce. Memory fills our dreams with people and places of the past. Yet experience in general seems more basic than memory. For one thing, there is a lot of novel experience in life, especially in the early years but continuing right through to the end. Moreover, when we recognize things or feel that our surroundings are familiar, it seems to me that it is only the sense of recognition or familiarity that depends on memory, not the whole experience in all its rich detail. In other words, memory seasons large tracts of experience that would exist in a memory-independent form without it.

<>

The following passages by Christof Koch (my emphasis throughout) illustrate a crucially ambiguous usage of the phrase "sufficient for":

These are the smallest set of brain mechanisms and events *sufficient for* some specific conscious feeling. (Koch 2004, xv)

The goal is to discover the minimal set of neuronal events and mechanisms jointly *sufficient for* a specific conscious percept. (Koch 2004, 16)

Much neural activity at any one time does not correlate with subjective states yet can still influence behavior. What is the difference between this and the activity that is *sufficient for* consciousness? (Koch 2004, 21–22)

If the input is more sustained or is boosted by a top-down attentional bias, on the other hand, it sets up long-lasting reverbatory activity that is powerful enough to generate the coalitions *sufficient for* conscious perception. (Koch 2004, 229)

The question I focus on is, What are the minimal neuronal mechanisms jointly *sufficient for* a specific conscious percept? (Koch 2004, 315)

In such passages, "sufficient for x" could mean either "sufficient to cause x" or "sufficient to enable one to reliably infer the existence of x." According to the quadrants view, we will never be able to specify properties that are sufficient to cause experience in general or particular experiential properties, because remote properties are always part of the mix. However, as researchers discover increasingly accurate and detailed correlations between features of experience and outwardly observable properties that play a role in producing them, it is possible that they will be able to specify outwardly observable properties from which one can reliably infer the existence of particular experiential properties. This is possible because the set of outwardly observable properties belongs to the indissoluble causally sufficient property package that also includes remote properties. If certain outwardly observable properties are present, then so are certain remote properties, and together all these properties produce certain experiential properties. This situation permits a reliable inference from the outwardly observable properties to the experiential properties; the outwardly observable properties are a sufficient basis from which to infer the existence of certain experiential properties but they are not sufficient to bring the experiential properties into existence.

This ambiguous usage of the phrase "sufficient for" helps to sustain the illusion that inferential sufficiency is equivalent to causal sufficiency. This illusion in turn supports the false hope that scientists will one day discover a set of properties that is sufficient to produce a given experiential property.

<>

John Searle is the victim of this illusion when he writes:

> Suppose we actually had an account of the neurophysiological processes in the brain that cause consciousness.... The knowledge of lawlike causal relations will give us all of the causal necessity we need. Indeed, we already have the beginnings of such lawlike relations.... standard textbooks of neurophysiology routinely explain, for example, the similarities and differences between how cats see things and how humans see things. *There is no question* that certain sorts of neurophysiological similarities and differences are *causally sufficient for* certain sorts of similarities and differences in visual experiences. (Searle 1992, 103; my emphasis)

Of special note here is Searle's extreme confidence in this analysis, expressed by the words "There is no question." The properties that Searle is so certain are causally sufficient are actually only inferentially sufficient, according to the quadrants view. They are the outward profile of a causally sufficient property package that also includes remote properties. In the same vein, Searle writes:

> Because the neurophysiological facts are *always causally sufficient* for any set of mental facts, *someone with perfect causal knowledge* might be able to make the inference from the neurophysiological to the intentional at least in those few cases where there is a lawlike connection between the facts specified in neural terms and the facts specified in intentional terms. (Searle 1992, 158–159; my emphasis)

In fact, the neurophysiological facts are *never causally sufficient* for a set of mental facts. Rather, they belong to the outward profile of a causally sufficient property package that also includes remote properties. It is quite possible that human beings will acquire the ability to infer the existence of specific mental facts from the existence of specific neurophysiological facts, but this will not be because they have perfect causal knowledge of the situation. It will be because they have sufficient causal knowledge of the situation to support the inference. Knowledge of outwardly observable properties can be sufficient to support the inference because the outwardly observable properties signify the presence of the complete, causally sufficient property package, which also includes remote properties. The "someone with perfect causal knowledge" that Searle imagines is impossible because the causation involves remote properties. I discuss other aspects of Searle's view in chapters 13 and 22.

<>

In *Consciousness and the Brain*, Stanislas Dehaene makes the following striking statements:

> Thanks to brain imaging, the mystery of consciousness has finally been cracked open. (Dehaene 2014, 117)

> In summary, neurophysiology has now cracked wide open the mystery box of conscious experience. (Dehaene 2014, 150)

If these statements are read as assertions of fact, there is nothing in Dehaene's book to support them. They appear to be overheated expressions of the author's enthusiasm for his research. Dehaene's research has the familiar goal of discovering correlations between an experimental subject's experiential states or transitions from one experiential state to another and concurrent outward observations of the person's body. The outward observations include EEG recordings from the person's scalp and functional MRI images of large regions of the person's brain. The discovery of this sort of correlation is not mystery-cracking.

In the following passage, Dehaene says that he is studying something more than "mere correlates" of consciousness, namely "genuine signatures" of consciousness that "encode" experience:

> These observations point to an all-important conclusion: we must learn to distinguish the mere *correlates of consciousness* from the genuine *signatures of consciousness*. Although the quest for the brain mechanisms of conscious experience is often described as a search for neural correlates of consciousness, this phrase is inadequate. Correlation is not causation, and a mere correlate is therefore insufficient. Too many brain events correlate with conscious perception—including, as we just saw, fluctuations that precede the stimulus itself and thus cannot logically be considered as coding for it. What we are

looking for is not just any statistical relation between brain activity and conscious perception, but a systematic signature of consciousness, which is present whenever conscious perception occurs and absent whenever it does not, and which encodes the full subjective experience that a person reports. (Dehaene 2014, 142; his emphasis)

However, the distinction that this passage describes is not between correlates and something more significant than correlates; it is between loose correlates and close correlates. By using the terms "signature" and "code" instead of "correlate," Dehaene puts the fact of close correlation in a sexy dress—another example of his enthusiasm outrunning his science.

Eventually, Dehaene admits that his laboratory work does not answer any of the questions that are usually associated with the phrase "the mystery of consciousness":

The discovery of signatures of consciousness is a major advance, but these brain waves and neuronal spikes do not explain what consciousness is or why it occurs. Why should late neuronal firing, cortical ignition, and brain-scale synchrony ever create a subjective state of mind? How do these brain events, however complex, elicit a mental experience? Why should the firing of neurons in brain area V4 elicit a perception of color, and those in area V5 a sense of motion? Although neuroscience has identified many empirical correspondences between brain activity and mental life, the conceptual chasm between brain and mind seems as broad as ever. (Dehaene 2014, 161–162)

He then outlines a plan for answering these questions:

Only mathematical theory can explain how the mental reduces to the neural. Neuroscience needs a series of bridging laws, analogous to the Maxwell-Boltzmann theory of gases, that connect one domain with the other. (Dehaene 2014, 163)

Like most writers, Dehaene consistently uses the word "neural" in an outwardly skewed sense. According to the quadrants view, the mental does not "reduce" to outwardly observable properties of the nervous system, so any attempt to explain *how* it reduces is misguided and futile. The salute to "mathematical theory" in this passage is hollow because it is obscure how mathematical theory could even engage with the stupendous variety of features of sensory, emotional, and thinking experience that Dehaene hopes to "reduce" to outwardly observable properties of the nervous system. The next mention of mathematical theory comes when Dehaene says that he seeks "a more sophisticated mathematical theory of how neural networks operate." His research team is pursuing this theory by means of "computer simulations of neural networks" (Dehaene 2014, 181). A key point that Dehaene does not make clear is that the computer simulations he has developed are exclusively simulations of outwardly observable properties of the nervous system such as firing frequencies and traffic patterns. Such simulations do not involve the experiential properties that are the crux of the theoretical problem. Dehaene may be confused about this. He misleadingly writes:

> Our model thus captured changes in the state of consciousness—the switch from an unconscious to a conscious brain. (Dehaene 2014, 185)

In fact, the only changes his model captures are changes in outwardly observable properties, some of which have been shown by his laboratory work to be *closely correlated with* experiential changes reported by human subjects. It is easy to believe that mathematical theory and computer simulations can shed some light on neural firing frequencies and traffic patterns, but this is no more a theoretical "bridge" to experiential properties than a riverside footpath is a bridge over the river.

Touching briefly on Chalmers's "hard problem," Dehaene makes the following bold prediction:

> Once our intuition is educated by cognitive neuroscience and computer simulations, Chalmers's hard problem will evaporate.... the science of consciousness will keep *eating away* at the hard problem until it vanishes. (Dehaene 2014, 262; my emphasis)

This use of the metaphor of "eating away" is interesting as well as ironic. Dehaene and his fellow neuroscientists are discovering more and more of the outwardly observable properties of the brain that play a role in generating experiential properties. In this sense, it seems fair to say that their work is eating away at the problem of explaining how experiential properties are generated. But this eating away is bounded in such a way that it is not approaching a solution to the problem. It is like a goat eating away at the grass inside a locked enclosure. Without the causally relevant remote properties, which are undiscoverable, the accumulation of more and more causally relevant outwardly observable properties merely sharpens the problem definition.

Dehaene and his colleagues are engaged in valid and valuable scientific work. An excellent example of the value of their work is Dehaene's description of the use of certain "signatures" of consciousness to determine whether brain-damaged patients who cannot communicate are comatose, "minimally conscious," or "locked-in" (fully conscious but totally paralyzed). Important treatment decisions hinge on such determinations. Neuroscientists should keep the "baby" of sound scientific research and throw out the "bathwater" of extravagant claims and futile ambitions.

<>

I conclude this chapter with criticisms of several grand statements concerning the scope of scientific research in general.

In his book *The "Mental" and the "Physical"*, Herbert Feigl advocates a so-called identity theory, which I discuss in chapter 21. Laying some groundwork for his theory, Feigl writes:

> Now, I think it is an essential aspect of the basic working program and of the working hypotheses of science that there is nothing in existence which would in principle escape intersubjective confirmation. Allowances have already been made for the (sometimes) insuperable practical difficulties of even the most incomplete and indirect confirmations. But the optimistic outlook that inspires the advance of science and informs its heuristic principles, does not tolerate the (objectively) unknowable or "un-get-at-able." (Feigl 1967, 33–34)

This passage puts forward the erroneous notion that the conduct of scientific research is somehow dependent on the belief that the potential scope of scientific discovery is nothing less than the whole of reality. In fact, people who conduct scientific research are simply *learning what they can* about the world through experiment, observation, and analysis. The success of their research does not depend on any "working hypotheses" or "optimistic outlook" of the sort that Feigl describes. All it depends on is the skilled performance of the relevant activities—experiment, observation, analysis. Feigl says nothing to justify the working hypotheses and optimistic outlook that he speaks of, and according to the quadrants view they are not justified. The remote properties of electronuclear fabric, as well as the experiential properties of animals that have no capacity for inward observation, are unknowable and "un-get-at-able." This limits the potential scope of scientific discovery in a certain respect, but it does not interfere with the conduct of scientific research. You don't need to assume that you can drive your car everywhere on the surface of the earth in order to drive around town. Similarly, people do not need to assume that they can get at every aspect of things through scientific research in order to get at many aspects of things in this way.

In *Consciousness: Creeping Up on the Hard Problem*, Jeffrey Gray writes:

> Physics aims to give a complete and completely unified account of the entire universe. Thus it cannot rest easy with a set of natural phenomena, such as those of conscious experience, which resist physical measurement and explanation. So either consciousness must be made to fit contemporary physics or physics itself must change to accommodate consciousness. (Gray 2004, 124–125)

Although some physicists may have the goal of developing "a complete and completely unified account of the entire universe," most physicists have much more modest ambitions. Gray is therefore wrong to say that this is a goal of the field of physics. As for the goal itself, it is futile according to the quadrants view, because an account of the universe cannot be complete without remote properties, and particular remote properties cannot find their way into any account. One cannot construct such an all-encompassing account by fitting consciousness to physics or by changing physics to accommodate consciousness. One cannot construct such an account at all.

In *The Quest for Consciousness*, Christof Koch writes:

> The first-person perspective, feelings, qualia, awareness, phenomenal experiences—call it what you want—are real

phenomena that arise out of certain privileged brain processes. They make up the landscape of conscious life: the deep red of a sunset over the Pacific Ocean, the fragrance of a rose, the searing anger that wells up at seeing an abused dog, the memory of the exploding space shuttle Challenger on live TV. Science's ability to comprehend the universe will be limited unless and until it can explain how certain physical systems can be sufficient for such subjective states. (Koch 2004, 315)

Science's ability to comprehend the universe *is* limited, according to the quadrants view. The reason is that science depends on observation and the electronuclear fabric of which things are made has remote properties that are inaccessible by observation, both outward and inward. Since experience is the joint product of outwardly observable properties and remote properties, one aspect of this limitation of science is an inability to "explain how certain physical systems can be sufficient for such subjective states." Koch returns to this theme a few pages later:

What matters from a metaphysical point of view is whether neuroscience can successfully move beyond correlation to causation. Science seeks a causal chain of events that leads from neural activity to subjective percept; a theory that accounts for what organisms under what conditions generate subjective feelings, what purpose these serve, and how they came about.

If such a theory can be formulated—a big if—without resorting to new ontological entities that can't be objectively defined and measured, then the scientific endeavor, dating back to the Renaissance, will have risen to its last great challenge. Humanity will have a closed-form, quantitative account of how mind arises out of matter. (Koch 2004, 326)

According to the quadrants view, science cannot provide the "closed-form, quantitative account" that Koch seeks because experience is jointly produced by outwardly observable properties (which are the bread-and-butter of science) and remote properties (which are beyond science). Humanity has access to only the outwardly observable portion of this collaboration.

Francis Crick concludes his book *The Astonishing Hypothesis* with the following rousing paragraph:

> How it will all turn out remains to be seen. The Astonishing Hypothesis may be proved correct. Alternatively, some view closer to the religious one may become more plausible. There is always a third possibility: that the facts support a new, alternative way of looking at the mind-brain problem that is significantly different from the rather crude materialistic view many neuroscientists hold today and also from the religious point of view. Only time, and much further scientific work, will enable us to decide. Whatever the answer, the only sensible way to arrive at it is through detailed scientific research. All other approaches are little more than whistling to keep our courage up. Man is endowed with a relentless curiosity about the world. We cannot be satisfied forever by the guesses of yesterday, however much the charms of tradition and ritual may, for a time, lull our doubts about their validity. We must hammer away until we have forged a clear and valid picture not only of this vast universe in which we live but also of our very selves. (Crick 1995, 262–263)

Crick is right to admit possibilities other than "the rather crude materialistic view many neuroscientists hold today" and "the religious point of view." But he is wrong to claim that the only sensible approach to this question is "further scientific work" and "detailed scientific research." He mistakenly equates rigorous inquiry with scientific research. There are good reasons to think that the "clear and valid picture" that Crick seeks includes the quadrants view, which is not a scientific finding. Let us by all means continue to "hammer away" scientifically at the body's electronuclear fabric, but let us not presuppose that scientific research can fathom it without remainder. Hammering away scientifically in the hope of understanding how our electronuclear fabric generates experience is like running in a hamster wheel in the hope of escaping from the cage.

12

The Quadrants and Panpsychism

There are many forms of panpsychism. The quadrants view is not one of them. According to the quadrants view, all experience as well as everything else that people normally call mental or psychological is generated within complex electronuclear systems of the kind that we call animals. It could be that complex electronuclear systems of a somewhat different kind, such as extraterrestrial life forms or futuristic robots, also have what it takes to generate experience, but most familiar electronuclear objects do not.

This chapter focuses on the form of panpsychism that is sometimes called panexperientialism. The central claim of panexperientialism is that extremely tiny objects such as electrons and molecular nuclei have experiential properties. This is the case, panexperientialists maintain, regardless of context or conditions: electrons have experiential properties whether they are in a living animal, a cement sidewalk, a snowball, or the blazing center of the sun. In chapter 7 I assumed that panexperientialism is wrong. In this chapter I explain why I consider that to be a reasonable assumption. I also note some key similarities and differences between panexperientialism and the quadrants view.

<>

The following sampling of quotations from various writers gives a sense of how panexperientialists imagine the experience that an electron has:

primordial dust of consciousness (de Chardin 1959, 73)

some sort of rudimentary psyche (de Chardin 1959, 77)

a sort of rudimentary consciousness (de Chardin 1959, 89)

at least some trace of experience (de Quincey 2002, 183)

its own low-level form of experience (de Quincey 2002, 200)

not altogether foreign to the feelings in our consciousness (Eddington 1958, 276)

primitive emotions, appetites and purposes (Griffin 1998, 137)

a slight appetition (Griffin 1998, 153)

a rudimentary mental pole (Griffin 1998, 153)

an aim at the future (however limited) (Griffin 1998, 153)

sensation and will (though, naturally, of the lowest grade) (Haeckel 1901, 224)

a rudimentary form of sensation and will (Haeckel 1901, 229)

feeling, perceiving, remembering, desiring, liking and disliking...in *some* form, however primitive and simple, however odd or strange, when compared to our human forms (Hartshorne 1978, 90; his emphasis)

very naïve purposes, desires, or feelings (Hartshorne 1978, 92)

feelings as different from ours as an atom is from our bodies (Hartshorne 1978, 95)

something in principle analogous to such feeling-qualities, though in detail doubtless very different (Hartshorne 1968, 178) some kind of qualitative character very alien to us (Rosenberg 2004, 94)

an experience that has a character in some very abstract sense *like* that of our experience but *specifically* unimaginable by us and unlike our own qualia (Rosenberg 2004, 95; his emphasis)

flashes of extraordinarily simple and brief feeling, like fireflies quietly flickering in the night (Rosenberg 2004, 96)

pure feeling too simple to support anything worthy of the name "consciousness" (Rosenberg 2004, 248)

somehow experiential in its essential and fundamental nature, however primitively or strangely or (to us) incomprehensibly (Strawson 2006, 24)

Two themes run through these descriptions. First, electron experience is much simpler than human experience. It is described as primordial, rudimentary, a trace, a low-level form, of the lowest grade, primitive, slight, limited, extraordinarily simple, and naïve. Second, we are not in a position to say anything definite about electron experience. It is described as odd, strange, very different, very alien, unimaginable, and incomprehensible. This indefiniteness is also indicated by the extensive use of "some" phrases: some sort, at least some trace, something more general, in some form, something in principle analogous, some kind of qualitative character, and somehow experiential.

These descriptions of simple and strange electron experience remind me of the folk medicine practice of homeopathy (Park 2000, 52–58; Brooks 2008, 181–202). Homeopathy combines two ideas. One is the paradoxical notion that you can cure a symptom with a substance that under normal circumstances causes that symptom, or at least a symptom very similar to it. For example, you can cure diaper rash with an extract from the poison ivy plant. The other idea is repeated dilution. In a typical dilution protocol, you start by making a 1% solution of the active ingredient, using either water or alcohol as the solvent. You then discard 99% of this solution, dilute what remains a hundredfold, discard 99% of that solution, and continue in this manner through 30 iterations. A simple calculation shows that after the last iteration there is less than a one-in-a-million chance that even one molecule of the active ingredient remains. The homeopathic substance, carefully selected at the start, is absent at the end.

Much as believers in homeopathic medicine use extreme dilution to escape the absurdity of treating diaper rash with poison ivy, believers in panexperientialism use a kind of extreme conceptual dilution to avoid an absurd attribution of experience to electrons. It would be bizarre to maintain that electrons have visual experience even though they have no eyes, or that they have feelings of thirst even though they are many orders of magnitude smaller than a water molecule. The idea that electrons have any specific and definite feature of sensory, emotional, or thought experience is absurd. To attribute experience to electrons without absurdity, one has to be very vague and dilute about it, as all the quoted passages are. The worry is that this avoidance of specifics undercuts the claim that electrons have experience, much as the homeopathic dilution protocol eliminates the active ingredient. If "experience" is an umbrella term for a variety of specific features, you can't get rid of every specific feature and still have experience left. This is not a conclusive argument against panexperientialism, but there is clearly a tension in maintaining that electrons have properties that are very, very different from familiar features of experience and yet sufficiently like familiar features of experience to be experiential.

One advocate of panexperientialism who has worried about this question is Gregg Rosenberg. He thinks the idea of electron experience can make sense because of "the open-ended character of our concept of experience," which he describes, in part, as follows:

> The privacy of consciousness forces us to build in a kind of tolerance for alien experiences and feelings: A manta ray sensing the electromagnetic structures on the ocean floor may experience qualities we could never imagine. We also have to allow that simpler and simpler organisms may have experiences of simpler and simpler kinds, as well as alien kinds. So the open-ended character of the concept requires us to accept that there could be experiences both very alien to, and much simpler than, any we can imagine. (Rosenberg 2004, 94)

I think this is well said. But this descent through a sequence of ever smaller and simpler organisms leaves us a long way from electrons. A typical bacterium with a diameter of one micron, arguably already much too simple to have any experience, contains on the order of a trillion (10^{12}) molecular nuclei and several trillion electrons. There may be a concept of experience that is open-ended, but there has to be something to distinguish it from the complete openness of a logical variable. Otherwise the statement "Electrons have experience" degenerates into the vacuous "Electrons have x."

<>

Let's assume for the sake of discussion that there is a way to understand the statement "Electrons have experience" such that it is neither vacuous nor absurd. Let's also assume that all electronuclear objects have outwardly observable properties and that no experiential property is outwardly observable. Panexperientialist writers do not explicitly make these last two claims, but I think that many of them would find these claims acceptable. Given these assumptions, panexperientialism and the quadrants view have the following significant similarity. According to both views, all electronuclear objects have properties that are not outwardly observable. According to both views, in other words, all electronuclear objects are propertywise plumper than their outward profiles. Coining a term, one could say that panexperientialism and the quadrants view are both forms of electronuclear panplumpism. In this respect, both of these views stand opposed to everyday outwardism, physical-science materialism, and physical-science physicalism regarding familiar electronuclear objects.

These two forms of panplumpism differ in the following significant respect. According to panexperientialism, all the not-outwardly-observable plumpness is experiential. According to the quadrants view, most electronuclear objects have only remote plumpness, with experiential plumpness being confined to living animals. One can, incidentally, imagine a third form of panplumpism according to which all electronuclear objects have both experiential plumpness and remote plumpness. However, everything I have to say about panexperientialism-without-remote-properties also applies to panexperientialism-with-remote-properties, so there is no need to give this third form of panplumpism special attention.

An important part of the quadrants view is the two-quadrant foundation hypothesis: the experiential properties of animals are joint products of outwardly observable properties and remote properties that are co-present in the indissoluble property packages of electronuclear objects. Panexperientialist writers do not present a complete alternative to this hypothesis. They all agree that the experiential properties of electrons and other very tiny things have some sort of causal or constitutional relation to the experiential properties of human beings and other living animals, but beyond this they say little.

Some writers seem to imagine the generation of more elaborate experience from simpler experience by a process that involves only experiential properties. Indeed, some believe that the world is predominantly or even exclusively experiential. One writer who can be read in this way is Eddington. The phrase used by Eddington that I quoted earlier in this chapter is part of the following passage:

> To put the conclusion crudely—the stuff of the world is mind-stuff.…. The mind-stuff of the world is, of course, something more general than our individual conscious minds; but we may think of its nature as not altogether foreign to the feelings in our consciousness. (Eddington 1958, 276)

He maintains that this ubiquitous mind-stuff is what we encounter through outward observation:

> Our view is practically that urged in 1875 by W. K. Clifford—
> "The succession of feelings which constitutes a man's consciousness is the reality which produces in our minds the perception of the motions of his brain."
> That is to say, that which the man himself knows as a succession of feelings is the reality which when probed by the appliances of an outside investigator affects their readings in such a way that it is identified as a configuration of brain-matter. (Eddington 1958, 278)

This remarkable passage raises a host of questions that Eddington does not address. How does it happen that visual observation of a succession of feelings seems to reveal neurons and molecules? If visual observation of something produces such a radical transformation, how can it have any cognitive value? How does it happen that visual observation of the center of a person's head yields results of very similar character whether the observed person is having a conscious succession of feelings at the time or is under general anesthesia? Eddington's mind-stuff view has extra problems that a panexperientialist can avoid by saying that the stuff of the world is complex stuff that has experiential properties in addition to its outwardly observable properties.

To design a form of panexperientialism that is as similar to the quadrants view as possible, one needs a structurally analogous counterpart of the two-quadrant foundation hypothesis. This counterpart hypothesis would be that the experiential properties of living animals are joint products of certain strange and simple experiential properties and certain outwardly observable properties that are co-present in the indissoluble property packages of electronuclear objects. To my knowledge, no panexperientialist writer has ever advanced this hypothesis. However, many of them say things that are suggestive of it. For example, panexperientialists generally draw a distinction between objects that have their own experience, such as people and pandas, and objects that don't have any experience of their own, such as rocks and chairs. Various terms have been coined to designate the class of objects that have their own experience, including "natural individuals," "genuine individuals," and "active singulars." Objects that do not have their own experience are "mere aggregates." How does it happen that there are these two classes of objects, given that all members of both classes are made of electrons and protons that, according to panexperientialism, have their own experience? It is open to a panexperientialist to note that experience-having animals have much more outwardly observable complexity than "mere aggregates" do, and to theorize that this outwardly observable complexity plays a role in generating the "advanced" experiential properties that are characteristic of animals. This is not what panexperientialists typically say; instead, they seem to imagine a purely experiential process that turns the strange and simple experience of electrons into the elaborate and familiar experience of human beings. But I think a panexperientialist could appeal to outwardly observable properties in this way without ceasing to be a panexperientialist.

<>

If these are the significant similarities and differences between panexperientialism and the quadrants view, which of these two views is more plausible?

Advocates of panexperientialism have constructed a large number of fallacious arguments for it. Following are eight of them.

Argument 1. As part of an exposition of the panpsychist view of Alfred North Whitehead, David Ray Griffin writes:

> Furthermore, once we have fully accepted the idea that our own experience is fully natural, therefore an (especially high-grade) example of natural events generally, we can generalize, saying that this twofold mode of existence must be true of the interactions within the body generally. Furthermore, realizing that the body is simply one more part of nature, we can generalize even further, saying that this twofold mode of existence must apply universally. (Griffin 1998, 148)

This passage suggests that "generalizing" experience from human beings to all of nature has some sort of logical legitimacy, but in fact this is a wanton maneuver. We live in a highly varied world. Many processes are fully natural yet specific to systems of a certain kind. Nuclear fusion takes place in stars, but not in rocks or plants. Photosynthesis takes place in plants, but not in rocks or stars. It is logically possible as well as plausible that experience is the product of special processes that take place in animals but not in plants, rocks, or stars. Griffin gives no reason to think otherwise.

Argument 2. Charles Hartshorne argues that the truth of panpsychism can be inferred from the reality of time:

> The truth seems to be rather that the idea of time is unintelligible unless panpsychism is true. For the only way in which we can conceive the unity of the different aspects of time—past, present, and future—is the way illustrated by our experiences of memory and anticipation. Without memory and anticipation "past" and "future" would be meaningless words; if nature does not remember and does not anticipate, we are forthwith at a loss how she has a past and a future. (Hartshorne 1968, 174)

This passage seems to depend on a gratuitously psychological conception of time according to which time is not merely a precondition of memory and anticipation, but somehow inseparable from them. Given a more usual conception of time, it is perfectly intelligible for an object to persist through time, undergoing various changes along the way, even though it is not capable of memory or anticipation.

Argument 3. Hartshorne also argues that the truth of panpsychism can be inferred from the nature of causation:

> In the broadest behavioral sense, remembering is taking account in present action of past events (experience) within the individual in question; perceiving is taking account in present action of past events in the environment. Even atoms take at least the immediate past into account; for if they did not there could be no causal account of their behavior. Therefore ... they either remember or perceive or both. (Hartshorne 1978, 94)

The problem here is an equivocal use of the word "is." The first sentence has some plausibility if it is understood as noting certain characteristics of remembering and perceiving. However, Hartshorne goes on to use this sentence as if it says that remembering and perceiving are fully defined by these characteristics. An argument of the same form and equal merit is "Quiche is food; amoebas need food; therefore amoebas need quiche."

Argument 4. Peter Ells arrives at panpsychism through a series of definitions. First he defines "experiential existence" and "empirical existence":

> **Definition:** to *exist experientially* is to exist as a mind that possesses qualia or qualitative experiences. Anything that has experiential existence is an *experiential being.* (Ells 2011, 55; his emphasis)

> **Definition:** to *empirically exist* is to have the power to cause (directly or indirectly in any manner) systematic, intersubjectively consistent regularities in the percepts of experiential beings. (Ells 2011, 56; his emphasis)

Then he uses these definitions to construct further definitions:

> **Definition:** an *experiential entity* is a unitary entity with both experiential existence and empirical existence.
> **Definition:** To *actually or concretely exist* (or just *exist* for short) is precisely to be an experiential entity, or to be composed of experiential entities.

> This latter definition is explicitly metaphysical. It aims
> to capture the concept of actual existence, and is therefore
> supposed to apply to all universes. (Ells 2011, 75–76; his
> emphasis)

In short, panpsychism is true because existence is experiential by definition! The trouble is that this last definition is a question-begging artifice that does not "capture the concept of actual existence" precisely because it makes existence logically dependent on experience. The dictionary definition of "existence" as "the state or fact of having being" is more to the point. It is not very illuminating, of course, but it has the merit of omitting extraneous elements.

Argument 5. Gregg Rosenberg's *A Place for Consciousness* includes a chapter entitled "The Probability of Panpsychism." What I see in this chapter is a set of arguments against certain alternatives to panpsychism. But other conceivable alternatives, including the quadrants view, are not considered. This kind of argument by elimination has no force unless you start with a set of alternatives that logically exhausts the possibilities.

Argument 6. David Skrbina says that if you reject panpsychism you face the "problem of 'drawing a line' somewhere, non-arbitrarily, between enminded and supposedly mindless objects" (2005, 250). He elaborates this as follows:

> If we grant that chimpanzees, say, or dogs, or dolphins
> possess even a *kind of* consciousness, or a *kind of* mind, then
> it seems that mind, as a *generalized phenomenon*, must exist
> in all things—because who could countenance drawing a line
> just *there*. (Skrbina 2005, 255; his emphasis)

This passage overlooks the basic fact that evolution produces large changes through the gradual accumulation of tiny changes. Every individual living thing is extremely similar to its parents. Accordingly, there is no place to draw a nonarbitrary line that marks the first occurrence of any characteristic of a living thing. For example, there is no nonarbitrary way to designate the first occurrence of a flower or the first occurrence of a bone, but it would be absurd to conclude from this that flowers and bones have always existed everywhere. One kind of thing morphs into another kind of thing through a series of nameless intermediate forms. Similarly, electronuclear systems that have no experience could have evolved into electronuclear systems that have experience through a series of subtly different quasi-experiential systems. There is no place to draw a nonarbitrary line, but the accumulation of tiny changes can be momentous nonetheless.

Argument 7. This argument resembles Argument 6 in that it focuses on the evolutionary progression from electrons and simple molecules to human beings. According to Sewall Wright:

> Emergence of mind from no mind at all is sheer magic.... mind cannot arise from nothing. (Wright 1978, 82)

Similarly, Christian de Quincey writes:

> The strong panexperientialist position states it is inconceivable that sentient, experiencing entities could evolve or emerge out of wholly insentient, nonexperiencing substance or events. If experience wasn't there to begin with, it couldn't emerge *ex nihilo.* This is the bottom line rationale for panexperientialism: Experience can only come from what has experience to begin with. (de Quincey 2002, 193)

This argument mischaracterizes the alternatives to panexperientialism. One can reject panexperientialism without believing that experience emerged "from nothing" or "*ex nihilo.*" Innumerable features have emerged in the course of evolution—photosynthesis, sexual reproduction, flight, intelligence, symbolic communication, and many more. The fact that there was a time when none of these features existed does not mean that they emerged "from nothing" or "*ex nihilo.*" They emerged from what was there before them. And it could be the same with experience.

Argument 8. Galen Strawson argues that "you can't make experience from something wholly nonexperiential" (2006, 29) because this would involve a "radically unintelligible transition" (2006, 28). I see three problems with Strawson's argument. First, it includes some obscure remarks about the word "intelligible" that suggest to me that Strawson may be giving this word an unusual meaning (2006, 15). Second, taking the word "intelligible" in an ordinary sense, it is not clear to me that the kind of panpsychism that Strawson advocates would make the world any more intelligible than the quadrants view does. I think that Strawson and other panpsychist writers create an illusion of intelligibility by throwing the word "experience" at everything. Where is the actual intelligibility? They give no explanation of how one feature of experience could emerge from a different feature of experience, let alone how familiar features of human experience could emerge from features of electron experience that are said to be odd, strange, unimaginable, and incomprehensible. Third and most importantly, again taking the word "intelligible" in an ordinary sense, I see no reason to accept Strawson's premise that there are no "radically unintelligible transitions" in the world. According to the quadrants view, remote properties and all the experience-generating processes in which they participate are radically unintelligible. I explain why this is a reasonable belief in chapter 15.

Fallacious arguments have also been made *against* panexperientialism. Here is one:

> Also, do the mental properties of the constituents of matter have any causal powers? Presumably they must if they are to give rise to mental states that do; but how is it, then, that particle physicists have not had to reckon with such causal powers in developing their theories of matter? If the mental properties of electrons bear upon how they will behave, then predictions about them will not be derivable from their physical properties alone: but we know this not to be the case—so the mental properties would have to be declared causally inefficacious. (McGinn 1982, 32)

This passage makes two mistakes. First, it assumes that the relation between prediction and causation is much closer than it actually is. People routinely make reliable predictions based on a partial understanding of the relevant causation or even without any understanding of the relevant causation. To make a reliable prediction, all you need is one property that signals the presence of a package of causally relevant properties. For example, when a traffic light turns orange you can reliably predict that it will soon turn red even if you have no understanding of how the device works. In like manner, physicists who make reliable predictions concerning the behavior of electrons might be unaware of some causally relevant properties, including mental properties. Second, this passage wrongly assumes that any electron property that plays a role in producing animal experience must also play a role in causing the electron behavior that physicists predict. It is logically possible that an electron has mental properties that play a role in producing animal experience but that do not play a role in repelling other electrons or in the absorption and emission of photons. In sum, the fact that physicists can make certain predictions about electrons on the sole basis of outwardly observable physical-science properties has no bearing whatsoever on the question of whether electrons have mental properties.

Notice incidentally that an argument that is structurally identical to this one can be made against the hypothesis that electrons have remote properties that play a role in producing animal experience. Such an argument against remote properties would make the same two mistakes. Remote properties of electrons could play a role in causing certain aspects of electron behavior that physicists are able to predict using only outwardly observable properties. And remote properties of electrons could play a role in producing animal experience without playing any role in causing the aspects of electron behavior that physicists are able to predict.

A fallacious argument does not support its conclusion, of course, but neither does it cast doubt on its conclusion. A statement can be as true as you please and someone can come along and make a fallacious argument for it. Thus, exposing the errors in all these fallacious arguments leaves the issue up in the air. What we need is good arguments! I am aware of two good arguments that favor the quadrants view over panexperientialism.

First is the homeopathy problem, which I discussed earlier in the chapter. I cannot convince myself that it makes sense to say that electrons have properties that are very, very different from familiar experiential properties and yet similar enough to them to be experiential properties. If this claim does not make sense, then it cannot be true. I admit that I also cannot convince myself that this claim does not make sense. But this state of uncertainty is not a neutral position. Being uncertain whether panexperientialism is even meaningful makes it impossible to consider it plausible.

Second, evidence of several types supports the idea that experiential properties exist only as part of an "animal package" that also includes a certain sort of neural complexity and certain behavioral capacities. Living animals with experiential properties have complex nervous systems, which seem to play an important role in producing the types of experience that animals have. In living animals, experiential properties seem always to have a behavior-guiding function that helps the animal meet its needs. Hunger drives animals to eat. Fear drives animals to flee. Sensory experience and thought help animals find their way around. In the course of evolution, moreover, experience, neuroanatomy, and behavior seem to have increased in complexity together. If experience, nervous systems, and the kind of behavioral flexibility that can benefit from experiential guidance make up a coordinated package, it is reasonable to expect these three things to be present or absent together. Electron experience would violate this expectation because it would have no supporting neural anatomy and no behavior-guiding function. Electron experience would be experience outside the one and only context that we have any reason to associate experience with.

No advocate of any form of panpsychism has compared panpsychism with the quadrants view, or even acknowledged the quadrants view as a competing hypothesis. Typically, advocates of panpsychism compare it, either explicitly or implicitly, with some other view that is also wrong. Such a comparison can make panpsychism seem attractive. My challenge to dedicated panpsychists of all stripes is to conduct a methodical head-to-head comparison of their view with the quadrants view. I doubt that any form of panpsychism can win this competition.

13

The Quadrants and John Searle

For the most part, John Searle uses the word "consciousness" rather than "experience." As I understand him, he uses the words "conscious" and "consciousness" in the broad sense that is coextensive with experience: an animal is conscious and has consciousness if and only it has some features of experience. In comparing Searle's view with the quadrants view, I use the word "consciousness" quite a bit, and always in this broad sense.

<>

A key characteristic of Searle's view is the distinction that he draws between two "modes of existence" or two "ontologies." He uses two adjectives for each side of this distinction—"subjective" and "first-person" for one side, "objective" and "third-person" for the other side. Notice that these are all terms that I do not use for reasons that are explained in chapter 5. But I mention them often in this chapter because they play such prominent roles in Searle's presentation of his view.

Searle describes his distinction between two "modes of existence" or "ontologies" in many places across his many books and articles. Here are four representative passages:

> Some entities, mountains for example, have an existence which is objective in the sense that it is does not depend on any subject. Others, pain for example, are subjective in that their existence depends on being felt by a subject. They have a first-person or subjective ontology. (Searle 1997, 113–114)

Some entities have a subjective mode of existence. Some have an objective mode of existence. So, for example, my present feeling of pain in my lower back is ontologically subjective in the sense that it only exists as experienced by me. In this sense, all conscious states are ontologically subjective, because they have to be experienced by a human or an animal subject in order to exist. In this respect, conscious states differ from, for example, mountains, waterfalls, or hydrogen atoms. Such entities have an objective mode of existence, because they do not have to be experienced by a human or animal subject in order to exist. (Searle 2002c, 23)

Some entities, such as pains, tickles, and itches, have a subjective mode of existence, in the sense that they exist only as experienced by a conscious subject. Others, such as mountains, molecules, and tectonic plates, have an objective mode of existence, in the sense that their existence does not depend on any consciousness. (Searle 2002b, 43)

Finally, and most important for our subject, conscious states are *subjective* in the sense that they are always experienced by a human or animal subject. Conscious states, therefore, have what we might call a "first-person ontology." That is, they exist only from the point of view of some agent or organism or animal or self that has them. Conscious states have a first-person mode of existence. Only as experienced by some agent—that is, by a "subject"—does a pain exist. Objective entities such as mountains have a third-person mode of existence. Their existence does not depend on being experienced by a subject. (Searle 1998, 42–43; his emphasis)

I find these passages obscure. I believe that one can distinguish two modes of observation or two modes of cognitive access—the outward and the inward. I believe that one can distinguish the set of properties that can be observed outwardly from the set of properties that can be observed inwardly. But I don't see that this distinction between two modes of observation lines up with a distinction between two "modes of existence" or two "ontologies." In my view, existence doesn't come in different modes; existence is existence. I and all my properties exist in the same plain way.

The obscurity of Searle's distinction is underscored by his use, in the last passage just quoted, of the seemingly nonsensical phrase that I discussed in connection with Nagel in chapter 8: "Conscious states ... *exist only from the point of view* of some agent or organism or self that has them" (my emphasis). Searle does not use this phrase often, but here it is again:

> Because of the qualitative character of consciousness, conscious states exist only when they are experienced by a human or animal subject. They have a type of subjectivity that I call ontological subjectivity. Another way to make this same point is to say that consciousness has a first-person ontology. It exists only as experienced by a human or animal subject and in that sense *it exists only from a first-person point of view.* (Searle 2004, 135; my emphasis)

For me, trying to make sense of the phrase "exist only from a point of view" is the intellectual equivalent of eating rocks.

What does Searle mean by "mode of existence" anyway? Given the flexibility of language, I am open to the possibility that different things have different modes of existence *in some sense*. But then what is the relevant sense? What thought am I supposed to have when I am told that pains and mountains have different modes of existence? Searle gives no explanation that I can find.

The word "ontology," as Searle uses it in these passages, is also puzzling. My dictionary gives two definitions for this word. First, there is a field of study called ontology, which is defined as "a branch of metaphysics concerned with the nature and relations of being." Second, a theory developed by someone working in the field of ontology is sometimes called *an ontology*. Neither of these definitions fits Searle's usage that some object or property "has an ontology." Presumably, then, Searle is using the word "ontology" in some novel, off-dictionary sense. What is this third sense? Searle gives no explanation that I can find.

An examination of the way in which Searle draws his distinction between the two "modes of existence" does not lessen the obscurity. The decisive factor for him seems to be whether or not the existence of something depends on being experienced. According to Searle, the "mode of existence" or "ontology" of anything that has this dependence is "subjective" and "first-person," whereas the "mode of existence" or "ontology" of anything that does not have this dependence is "objective" and "third-person." I can appreciate a distinction between dependence and independence, but I don't see how such a distinction gives rise to different "modes of existence" or "ontologies." Consider structurally parallel examples such as the following:

> The existence of rivers and lakes depends on rainfall, but the existence of mountains and deserts does not. So rivers and lakes have a rainfall-dependent mode of existence, whereas mountains and deserts have a rainfall-independent mode of existence.

> The existence of beer and wine depends on alcohol, but the existence of milk and coffee does not. So beer and wine have an alcoholic ontology, whereas milk and coffee have a nonalcoholic ontology.

> The existence of automobiles and computers depends on human manufacturing processes, but the existence of birds and trees does not. So automobiles and computers have a man-made mode of existence, whereas birds and trees have a non-man-made mode of existence.

> The existence of some islands depends on volcanos, but the existence of other islands does not. So some islands have a volcanic ontology, whereas other islands have a nonvolcanic ontology.

My impression is that Searle thinks he is using "mode of existence" and "ontology" in a sense that does not apply willy-nilly to every distinction between dependence and independence. Otherwise, he is using some very unusual terminology to do the same job that the word "dependence" does. But if this is so, he should be able to explain what the crucial difference is between dependence on being experienced and other cases of dependence. Here it is not enough to note that dependence on being experienced is unique; every case of dependence is unique in its own way. What needs to be shown is that dependence on being experienced is unique in a special way that makes the terms "mode of existence" and "ontology" uniquely appropriate for it. I don't think this can be done. I think the distinction between what does and what does not depend on being experienced is a real distinction, but I don't see anything about this distinction that warrants Searle's description of it as a distinction between two "modes of existence" or two "ontologies."

I see an interesting similarity here between Searle and Descartes. According to Descartes, inward observation gives us access to properties of a "thinking thing" that is different from all the "extended things" that we encounter through outward observation. According to Searle, inward observation gives us access to a "subjective mode of existence" that is different from the "objective mode of existence" of the objects that we encounter through outward observation. Thus, Descartes and Searle both affirm a cognition-independent distinction that underlies the cognitive distinction between outward observation and inward observation. For Descartes, the underlying distinction is between extended unthinking things and unextended thinking things, so-called substance dualism. For Searle, the underlying distinction is between two "modes of existence" or two "ontologies." To highlight this similarity, one could call Searle's view mode-of-existence dualism. According to the quadrants view, the world is not bifurcated in either of these ways. We are, and we live among, existing electronuclear objects that have existing properties.

Descartes's distinction gives rise to the famous question about how the extended unthinking thing and the thinking unextended thing connect and interact within a person. In a similar manner, Searle's distinction gives rise to a question about how a single animal can straddle the two modes of existence. To my knowledge, Searle never addresses this question. All his examples of the modes of existence are either purely "subjective" features of experience such as pains and itches or purely "objective" experience-free objects such as mountains and hydrogen atoms. He never says that there is a third category that contains double-mode-of-existence conscious animals, even though the existence of such a hybrid or mixed category is an immediate consequence of his distinction between two modes of existence.

How significant is this issue? I agree that Searle's distinction between what does and what does not depend on being experienced is a real one. Might it be that his description of this distinction in terms of contrasting "modes of existence" or "ontologies" is just a matter of awkward terminology that can be set aside, revealing a view that is substantively correct? It turns out that this is not the case. Searle's terminology of "modes of existence" and "ontologies" is bound up with a fundamental problem, to which I now turn.

<>

Two of the adjectives that Searle uses for his contrasting "modes of existence" are "first-person" and "third-person." As I noted in chapter 5, these adjectives are often used to mark a distinction between two modes of cognition, namely inward observation of your own experience and outward observation of whatever is at the far end of the spatial transmissions that strike your sense organs. Searle himself sometimes uses "first-person" and "third-person" in this way. Thus, Searle describes the distinction between what does and what does not depend on being experienced using adjectives that he also uses to distinguish two modes of cognition. This prompts the suspicion that he is inadvertently fusing these two distinctions together. There is considerable evidence that he is doing exactly this, and generating fallacious arguments as a result. Consider first the following passage:

> If we think of the world as consisting of particles, and those
> particles as organized into systems, and some of those systems
> as biological systems, and some of those biological systems as
> conscious, and consciousness as essentially subjective—then
> what is it that we are being asked to imagine when we imagine
> the subjectivity of consciousness? After all, all those other
> things we imagined—particles, systems, organisms, etc.—
> were completely objective. *In consequence*, they are equally
> accessible to all competent observers. (Searle 1992, 96; my
> emphasis)

Recall that, according to Searle's definition, "having an objective ontology" means existing independently of being experienced. This definition says nothing about being accessible to all competent observers. Lack of dependence on being experienced has no conceptual connection with being accessible to competent observers. The two ideas are unrelated. This is why there is a logical possibility that electronuclear objects have remote properties, which are neither experiential nor accessible to observers. Why then does Searle say that being "equally accessible to all competent observers" is a *consequence* of being "completely objective"? The reason, I believe, is that his complete conception of "having an objective ontology" includes the condition of being outwardly observable—"third-person" in a cognitive sense—even though this condition is not part of his stated definition of "having an objective ontology." His stated definition has only one condition—existing independently of being experienced. But he uses the term as if it stands for a conjunction of two logically independent conditions—existing independently of being experienced *and* having a nature that is wholly open to outward observation.

Much depends on which way the phrase "having an objective ontology" is understood. Searle lists many things that in his judgment "have an objective ontology." For example, in the passages quoted at the beginning of this chapter he mentions mountains, waterfalls, hydrogen atoms, molecules, and tectonic plates. If "having an objective ontology" stands for the single condition of existing independently of being experienced, then the case that these things "have an objective ontology" is strong. However, if "having an objective ontology" stands for a conjunction of two logically independent conditions—existing independently of being experienced *and* having a nature that is wholly open to outward observation—then the case that mountains, molecules, and so on "have an objective ontology" is extremely weak, if not nonexistent. According to the quadrants view, all electronuclear objects have a plenary nature that includes properties that are not outwardly observable. The outwardly observable properties of such objects might be said to "have an objective ontology" in the two-condition sense, but not the plenary objects. I have given several reasons why this is a plausible conjecture. It could be wrong, but without a reason to think it is wrong Searle has no case that mountains, molecules, and other plenary electronuclear objects "have an objective ontology" in the two-condition sense.

Here is another passage that contains a dubious use of the phrase "in consequence":

> What more can we say about this subjective mode of existence? Well, first it is essential to see that *in consequence* of its subjectivity, the pain is not equally accessible to any observer. Its existence, we might say, is a first-person existence. (Searle 1992, 94; my emphasis)

Recall that, according to Searle's definition, "having a subjective mode of existence" means being dependent on being experienced. This definition says nothing about not being accessible to observers. Dependence on being experienced has no conceptual connection whatsoever with not being accessible to observers. The two ideas are unrelated. According to the quadrants view, the properties of electronuclear objects that are not accessible to all observers include experiential properties, whose existence depends on being experienced, and also remote properties, whose existence does not depend on being

167

experienced. The reason that experiential properties and remote properties are not accessible to all observers is that they can have no evidentiary effects on the spatial transmissions. The fact that the experiential properties depend on being experienced is incidental; it does nothing to explain why these properties are not outwardly observable.

What these two passages suggest is that the distinction between what does and what does not depend on being experienced and the distinction between what is and what is not accessible to all observers are glued together in Searle's thinking in such a way that he treats them as a single distinction. The world according to Searle has two compartments. The objective, third-person compartment contains everything that exists independently of being experienced *and* is wholly open to outward observation. The subjective, first-person compartment contains everything whose existence depends on being experienced *and* is wholly closed to outward observation. The trouble with this analysis is that he gives no reason why the conjoined pairs of conditions should always be satisfied together. He therefore gives no reason why these two compartments should be exhaustive. Why mightn't electronuclear objects have properties that exist independently of being experienced and are closed to outward observation? Such properties would not belong to either of Searle's two "ontologies." The possibility of remote properties is locked out of Searle's thinking by his quasi-Cartesian dichotomous terminology.

It seems possible to me that one factor leading Searle to his two-condition sense of "objective ontology" is the ambiguity of the word "objective" as discussed in chapter 5. Searle discusses the ambiguity of "subjective" and "objective" in many of his books and articles. But he only discusses two meanings of each word, neglecting their use to mark the distinction between inward observation and outward observation. Fusing "objective" in the sense of "having reality independent of the mind" to "objective" in the sense of "outwardly observable" would suffice to produce Searle's two-condition notion of "objective ontology." Another possibility is that Searle is simply in the grip of outwardism regarding nonliving objects, and that his outwardism enters his writing in the form of the two-condition sense of "having an objective ontology." Perhaps both factors are at work. Perhaps the fact that there is a much-used English word that can mean either "having reality independent of the mind" or "accessible by all competent observers" reflects the grip of outwardism on those whose use of this word over the centuries has made it the semantic mish-mash that it is today.

<>

Searle's terminological lockout of remote properties plays a crucial role in his discussion of unconscious intentionality and unconscious aspectual shape. "Intentionality" is an unpleasant word that has become standard in discussions of the fact that thinking is always about something. Following is one of many passages in which Searle explains his use of this word:

> "Intentionality" is a technical term used by philosophers to refer to that capacity of the mind by which mental states refer to, or are about, or are of objects and states of affairs in the world other than themselves. So, for example, if I have a belief, it must be a belief that something is the case. If I have a desire, it must be a desire to do something or that something should happen. If I have a perception, I must at least take myself to be perceiving some object or state of affairs in the world. (Searle 2004, 28)

The aspectual shape of a thought consists of those aspects of the thing that is being thought about that are represented in the thought. Metaphorically speaking, it is the handle by which the thinker grasps what he is thinking about. One of Searle's standard examples of aspectual shape is thinking about water. In most cases, a person thinking about water thinks of it as liquid, transparent, and thirst-quenching, but not as having the chemical formula H_2O or as being prone to dissociate into H^+ and OH^- ions or as having any number of other properties that a chemist could tell you about. On the other hand, a chemist thinking about water might envision a hustle and bustle of H_2O molecules, H^+ ions, and OH^- ions, all vibrating, rotating, and jostling one another—a very different aspectual shape.

Concerning intentionality and aspectual shape understood in this way, Searle poses the following interesting question. When a person is completely unconscious—asleep and not dreaming, for example—in what sense, if any, do the multitudinous beliefs and memories stored in his brain have intentionality and aspectual shape? To make this question more concrete, I will use an example from my own life. Every now and then I find myself contemplating a certain vivid memory from age five or six. In this memory, I am standing on a straight, gently sloping asphalt walk in a playground in the neighborhood where my family lived at the time. The walk is wet from a rain that has just ended, and it is littered with the motionless bodies of innumerable dead earthworms. I am taken aback by this spectacle, and feel a sense of amazement

169

and incomprehension. Presumably this was the first time in my life that I had seen such a thing. My contemplation of this memory has intentionality; I am thinking about a certain playground, about rain, about earthworms, about death. It has aspectual shape; the shiny gray of the wet pavement and the curly brown bodies of the worms are part of the recollection, but countless details concerning asphalt walks and rain and earthworms are not. It also has an emotional tone, another common dimension of thinking that Searle could have added to his discussion. Searle's question concerns the form in which this memory has been stored in me over the many intervening years. In what sense, if any, does the continuously stored memory have intentionality and/or aspectual shape?

His answer to this question has two parts. The part of interest here is that intentionality and aspectual shape are not inherent characteristics of any stored belief or memory content. The other part is that stored belief and memory content can be said to have intentionality and aspectual shape in a derived and secondary sense, because it has the capacity to produce episodes of conscious contemplation that do have inherent intentionality and aspectual shape.

In support of this answer, Searle presents an interesting argument. To make the nature of this argument clear, I will begin by noting something that Searle does not say. He does not say that consciousness is part of the very definition of inherent intentionality and aspectual shape. Such a definition would make it logically impossible for unconsciously stored memory and belief content to have inherent intentionality and aspectual shape, so no argument would be necessary. Rather, his definitions leave open the logical possibility that intentionality and aspectual shape are inherent attributes of unconsciously stored belief and memory content. Whether this logical possibility is ever realized is therefore a question of fact. This is the question that his argument addresses.

Searle's argument, in essence, is that during periods of unconsciousness a person consists entirely of "objective ontology," which cannot support inherent intentionality or aspectual shape. Here is one passage in which he makes this argument:

> *But the ontology of unconscious mental states, at the time they are unconscious, consists entirely in the existence of purely neurophysiological phenomena.* Imagine that a man is in a sound dreamless sleep.... he believes that Denver is the capital of Colorado, Washington is the capital of the United States, etc. But *what fact about him makes it the case that he has these*

unconscious beliefs? Well, the only facts that could exist while he is completely unconscious are neurophysiological facts....there is simply nothing there except neurophysiological states and processes.

But now we seem to have a contradiction: The ontology of unconscious intentionality consists entirely in third-person, objective, neurophysiological phenomena, but all the same the states have an aspectual shape that cannot be constituted by such facts, because there is no aspectual shape at the level of neurons and synapses. (Searle 1992, 159; his emphasis)

The resolution of this seeming contradiction, Searle goes on to say, is that stored belief content has intentionality and aspectual shape in a derived and secondary sense, due to its role in producing conscious thinking that has inherent intentionality and aspectual shape.

Whatever force this argument appears to have derives from the ambiguity of "having an objective ontology." I agree that the brain of an unconscious person "has an objective ontology" in the one-condition sense that is stated in Searle's definition of this term: the brain of an unconscious person, including all its parts and properties, exists independently of being experienced. But this one-condition sense of "having an objective ontology" leaves open the possibility that stored belief and memory content has inherent intentionality and aspectual shape. To make his argument, Searle needs the two-condition sense of "having an objective ontology"; the claim that does the work is that there is nothing to the brain of an unconscious person except outwardly observable "neurophysiological facts" and "neurophysiological phenomena." But Searle gives no reason to believe this claim. He gives no reason to rule out the possibility that the brain of an unconscious person has remote properties, both fundamental and evolved, some of which could give inherent intentionality and aspectual shape to stored memory and belief content. All he gives us is an equivocal usage of "having an objective ontology," which hides the difference between the one-condition claim for which one can make a strong case and the two-condition claim that his argument requires.

Here are two more passages in which Searle makes this same argument:

Where intentionality is concerned, there is a crucial difference between wanting water and wanting H_2O, because intentionality requires an aspectual shape and the aspectual

171

shape of these two intentional states is different. But when an agent is unconscious there is no aspectual shape present in his mind. The only reality the system has when unconscious is that of a series of ontologically third-person phenomena, such as neuronal structures and neuron firings. What then makes the aspectual shape of the unconscious person different in the case of wanting water and wanting H_2O? The only answer to that is that the different unconscious states are capable of manifesting themselves as a conscious state. The conscious state of wanting water is different from the conscious state of wanting H_2O, and the potentiality of producing that conscious state must exist when the agent is unconscious. (Searle 2010a, 23–24)

Ask yourself, What fact corresponds to the claim that Jones unconsciously wants water? If the state is totally unconscious, then the fact must be purely neurobiological. But the neurobiological, under neurobiological description, has no aspectual shape. (Searle 2010b, 204)

In these passages, the phrases that smuggle in the second condition of complete accessibility to outward observation are "ontologically third-person phenomena," "purely neurobiological," and "the neurobiological, under neurobiological description." Again, Searle gives no reason to think that the plenary brain of an unconscious person is wholly accessible to outward observation. He gives no reason to rule out the possibility that it has remote properties—properties that are neither outwardly observable nor experiential—and he gives no reason to rule out the possibility that some of these remote properties have inherent intentionality and aspectual shape. All he does is glue the condition of existing independently of being experienced to the logically independent condition of being accessible to outward observation, under cover of the jargon of "having an objective ontology."

Here is yet another presentation of the same argument:

What is wrong with just saying of the processes in the brain that they are unconscious intentional states occurring right then and there as unconscious intentional states? Why do we have to go through this elaborate dispositional analysis where we say that the attribution of unconscious intentionality is like describing something as poison or bleach? The answer is that the neurobiology as such has no aspectual shape.... The neurophysiology, described in terms of synaptic strength and action potentials, knows nothing of aspectual shape. (Searle 2004, 247)

In this passage, the unsupported claim that the brain of an unconscious person is wholly open to outward observation lurks in the murky phrases "the neurobiology as such" and "the neurophysiology, described in terms of synaptic strength and action potentials." The phrases "as such" and "described in terms of" have nothing to do with the reality of a living brain.

This last passage also illustrates Searle's habit of using what are normally names of scientific disciplines—"neurobiology" and "neurophysiology" in this case—to refer to the parts of the world that scientists in those disciplines study. Here are some other passages in which this habit is on display:

Quite so, but in order to imagine such a world, you have to imagine a change in the laws of nature, a change in those laws by which physics and biology cause and realize consciousness. (Searle 1997, 173)

Throughout this book we have assumed that at any given instant the state of a person's consciousness was entirely determined by her neurobiology. (Searle 2004, 226)

Notice in this case the neurobiology is capable of causing the pain in a conscious form, even though when I am sound asleep I do not consciously feel any pain. (Searle 2004, 244)

Using the names of scientific disciplines in this way has the effect of obscuring the distinction between science and nature. It fosters the impression that the world is a "science object" whose nature coincides with the possible extent of scientific knowledge. Such a notion dovetails nicely with the idea that most of the world consists of "objective ontology" that is wholly accessible by outward observation. But it is without foundation. There is no reason to think that the relationship between nature and science is this cozy.

The topic of unconscious intentionality and unconscious aspectual shape is tricky in the following respect. Our only possible source of illustrative examples of inherent intentionality and inherent aspectual shape is conscious thinking. No one will ever be able to point to an example of inherent intentionality or aspectual shape that is stored in the form of remote properties. It is therefore impossible to think about inherent unconscious intentionality and inherent unconscious aspectual shape in a concrete way. For this reason, I am neutral on the question of whether unconsciously stored belief and memory content has inherent intentionality and inherent aspectual shape. What I advocate is the more general idea that unconsciously stored belief and memory content might well be in a format that includes both fundamental and evolved remote properties in addition to the outwardly observable properties that neurophysiologists or other brain scientists can discover. Recapitulating points made in chapter 7, the case for this idea is as follows. It is plausible that all electronuclear objects have remote properties. We know that people have many experiential properties and many outwardly observable properties that have been produced through evolution. This makes it plausible that people also have remote properties that have been produced through evolution; there is no reason why the remote quadrant should not participate in the evolutionary process. If there are evolved remote properties, then some of these evolved remote properties could be integral to the format of unconsciously stored belief and memory content. A reason to take this possibility seriously is that it would give the two-way traffic between conscious thinking and unconsciously stored belief and memory content a homogeneity that seems like a simple and efficient way for a brain to operate. If conscious content and unconsciously stored content exist in similar formats, then less format conversion would be required in forming new memories and making conscious use of old memories. Whether the brain actually works this way must remain a matter of speculation. The important point is that Searle's argument that the brain cannot work this way is fallacious because it depends on the equivocation between the one-condition sense and the two-condition sense of "having an objective ontology."

<>

Searle's terminological lockout of remote properties also plays a crucial role in his account of the generation of experience in a living animal. The gist of his account is that "subjective ontology" is entirely caused by "objective ontology," where "objective ontology" has the two-condition sense. What this means in practice is that every property that plays a role in producing any particular feature of experience is potentially accessible by scientific researchers. As I noted in chapter 11, Searle thus looks forward to the day when we will have complete explanations of how consciousness and various particular features of experience are produced. He grants that the problem is difficult, but he thinks that the history of science provides encouragement with its many examples of difficult problems that were solved:

> ... given our present scientific paradigms, it is not clear how consciousness could be caused by brain processes. But I see this as analogous to the following: within the explanatory apparatus of Newtonian mechanics, it was not clear how there could exist phenomena such as electromagnetism; within the explanatory apparatus of nineteenth-century chemistry, it was not clear how there could be a non-vitalistic, chemical explanation of life. That is, I see the problem as analogous to earlier apparently unsolvable problems in the history of science ...
>
> My own guess ... is that when we have a general theory of how brain processes cause consciousness, our sense that it is somehow arbitrary or mysterious will disappear. In the case of the heart, for example, it is clear how the heart pumps blood. Our understanding of the heart is such that we see the necessity. Given the contractions, it causes blood to flow through the arteries. What we so far lack for the brain is an analogous account of how the brain causes consciousness. But if we had such an account—a general causal account—then it seems to me that our sense of mystery and arbitrariness would disappear.
>
> ... It now seems mysterious to us that neuron firings in the thalamus should cause sensations of pain. And I am suggesting that a thorough neurobiological account of exactly how and why it happens would remove this sense of mystery. (Searle 2002c, 24–25)

Searle thinks that he has provided the correct conceptual framework within which the relevant scientific research must proceed:

Neuronal behavior causes consciousness. It may be possible to produce in systems other than neuronal systems, but so far we do not know how to do it. The basic way to explain how intentionality is possibility is to explain how conscious forms of intentionality are possible, and that makes our neurobiological explanation of intentionality dependent on our explanation of consciousness. However, there are many different specific forms that will require separate explanations. So, for example, perception is different from intentional action, though they are of course intertwined in all sorts of ways in our actual life. Both intentional action and perception will presumably require appealing to different sorts of explanatory mechanisms than is the case, for example, with thought processes and emotions. All of these are empirical questions that I leave to neurobiology and neuropsychology, and to cognitive science generally. The philosophical point is that the general relationship between the human reality and the basic reality is clear. The details have to be worked out by the special sciences. (Searle 2010a, 24)

All of this flows from Searle's equivocation on "having an objective ontology." Once you recognize that having no dependence on being experienced and being accessible to all competent observers are logically independent conditions, and thus that remote properties are a logical possibility, Searle's reasoning unravels. There is no reason to accept his "philosophical point" about "the general relationship between the human reality and the basic reality," and there is no reason to share his confidence that the march of science will produce explanations of the causal foundations of consciousness, intentionality, perception, thought, and emotion.

According to the quadrants view, the correct account of "the general relationship between the human reality and the basic reality" is that consciousness and all particular features of experience are jointly caused by outwardly observable properties that scientists can discover and remote properties that scientists cannot discover. Accordingly, there will never be a correct scientific explanation of the causation of consciousness or of any particular feature of experience. The best that scientists can achieve is a complete reckoning of the causally relevant outwardly observable properties. The result of continued scientific progress will be a growing and increasingly accurate table of correlations between features of experience and the outwardly observable properties that play a role in producing them. The causally crucial remote properties that always accompany the causally crucial outwardly observable properties will never be available for use in explanations. The case for this account has two mains parts. First is the case that all electronuclear objects have remote properties. The details are in chapter 7. Second is the case that if all electronuclear objects have remote properties, then it is very likely that some of these remote properties play a role in producing consciousness. The details are in chapter 10. This case is inconclusive, but strong. It is much stronger than Searle's case for his account, which rests on the equivocation between the one-condition sense and the two-condition sense of "has an objective ontology." Without this jargon and its terminological lockout of remote properties, Searle has no case at all.

Something that I have always liked about Searle's work is his forceful criticism of the many theorists who have sought to understand human beings in purely outward terms. Behaviorism, functionalism, and the most common forms of materialism and physicalism all fall in this category. I agree wholeheartedly with the following passage, in which Searle sums up his criticisms of these views:

> More than anything else, it is the neglect of consciousness that
> accounts for so much barrenness and sterility in psychology,
> philosophy of mind, and cognitive science. (Searle 1992, 227)

But my attitude is different when he attributes this barrenness and sterility to a "persistent failure to recognize and come to terms with the fact that the ontology of the mental is an irreducibly first-person ontology" (Searle 1992, 95). He is right to criticize the purely outward theories, but he is wrong to put in their place the scheme of the two "ontologies." His fundamental error is his belief that experiential properties are the whole of what these purely outward theories neglect. According to the quadrants view, the purely outward theories neglect not only the experiential properties of humans and other animals, but also the remote properties of all electronuclear objects, animals included. Searle shares an important neglect with the purely outward theories that he criticizes—neglect of remote properties.

<>

Another problem with Searle's view is his extensive use of the idea, criticized in chapter 2, that a human body contains a hierarchy of levels. In particular, he makes the strong claim that consciousness is located at a higher level than all of its causes. Here is one of the many passages in which he says this:

> Mental phenomena are caused by lower-level neuronal processes in human and animal brains and are themselves higher-level or macro features of those brains. (Searle 2002a, 70)

Some passages suggest the similar but even stronger claim that mental properties have a level all to themselves. In other words, all nonmental properties of a person are at a lower level than consciousness, whether they play a role in causing consciousness or not. Here is one example:

> I have said that the philosophical solution to the traditional mind-body problem is to point out that all of our conscious states are higher-level or systemic features of the brain, while being at the same time caused by lower-level microprocesses in the brain. *At the system level we have consciousness, intentionality, decisions, and intentions.* At the micro level we have neurons, synapses, and neurotransmitters. The features of the system level are caused by the behavior of the micro level elements. (Searle 2007, 58–59; my emphasis)

The implication of this passage seems to be that the dividing line between "subjective ontology" and "objective ontology" coincides with the dividing line between the top level of the body's hierarchical structure and the level immediately beneath it. Searle never says this explicitly. In fact, to my knowledge he never explicitly discusses the relation between his "ontologies" and his "levels." But his ontologies and his levels must be related in some way, and this is the relation that seems to have the most textual support.

There are two reasons to reject all of this. First, as I explained in chapter 2, there is no literal hierarchy of levels in the body. Therefore, claims about the "levels" occupied by consciousness and its causes cannot be literally true. Second, using the word "level" in a loose sense, there is no reason to think that consciousness occupies a level all to itself and that all its causes occupy lower levels. It seems entirely possible that some properties that play a role in causing consciousness occupy the same "high," "system" or "macro" level that consciousness occupies. The following passage written by Searle himself in a more circumspect mood mentions several possibilities:

> What is the right level for explaining consciousness? Is it the level of neurons and synapses, as most researchers seem to think, or do we have to go to higher functional levels such as neuron maps or whole clouds of neurons; or are all of these levels much too high so that we have to go below the level of neurons and synapses to the level of the microtubules? Or do we have to think much more globally in terms of Fourier transforms and holography? (Searle 2002b, 38)

Loosely speaking, neuron maps, clouds of neurons, Fourier transforms, and holography might easily be placed at the same "high," "system" or "macro" level as consciousness. In addition, it seems entirely possible that there are properties at the same "high," "system" or "macro" level as consciousness that have nothing in particular to do with producing consciousness. What about a person's posture, metabolic rate or blood pressure? What about the spongy character of brain tissue? Searle frequently likens the relationship between consciousness and its "lower level" causes to the relationship between the solidity of a large object and its "lower level" causes. If one accepts this analogy, then it seems to me that the brain's sponginess might easily be placed at the same "high," "system," or "macro" level as the brain's consciousness—loosely speaking.

Other writers have applied the word "level" to this subject in a variety of other ways. It is common, for example, to say that the processing of sensory input goes through a series of levels between a sense organ such as the eye or ear and the part of the brain that generates the resulting sensory experience. These are levels of increasing abstraction or sophistication in a processing hierarchy, not levels of increasing size in a structural hierarchy. Or again, there is a traditional idea associated with substance dualism that "the mind" is on a higher level than the body in the sense of being more august or estimable. The mind is magnificent, while the body is made of "mere matter." Then there is the familiar purely metaphorical usage of the word "level" that is exemplified by expressions such as "below the level of conscious experience" (Dehaene 2014, 59) and "below the level of our awareness" (Dehaene 2014, 79). The same metaphor informs the word "subconscious." It could be that some of Searle's confusing statements about levels result from mixing the idea of a structural hierarchy of levels with one or more of these other uses of the word. This is just a speculation. It would help to explain some of Searle's statements, but other explanations are also possible.

<>

I have said nothing in this chapter about Searle's account of the role of consciousness in the causation of human behavior. I discuss this aspect of his view in chapter 22, as part of my comprehensive discussion of the causation of human behavior.

<>

Searle labels his view "biological naturalism." Recall from chapter 8 that the two words "biological" and "natural" are ambiguous and subject to outward skewing. If both of these words are used without any outward skewing, "biological naturalism" might mean simply that consciousness is an aspect of nature and a distinctive property of certain living systems. Understood in this way, this term does not bother me. One could say that the quadrants view is a form of biological naturalism in this sense. I think Searle does use the word "naturalism" without outward skewing, but he seems to use the word "biological" both without and with outward skewing. In the following sentence, for example, "biological" seems to indicate only a connection with living things:

> Consciousness is above all a biological phenomenon, like digestion or photosynthesis. (Searle 1997, 175)

But when it comes to explaining the label "biological naturalism," Searle seems to associate "biological" with outward-observation-based biological science:

> ... my label for this view is "biological naturalism": "naturalism" because, on this view, the mind is part of nature, and "biological" because the mode of explanation of the existence of mental phenomena is biological—as opposed to, for example, computational, behavioral, social, or linguistic. (Searle 1998, 54)

According to the quadrants view, Searle is right that consciousness is a "biological phenomenon" in the sense of being an aspect of certain living things, but he is wrong that "the mode of explanation of the existence of mental phenomena is biological" because the role of remote properties in the generation of mental phenomena makes biological explanations, or any other explanations, of how they are generated impossible. He correctly rejects "computational, behavioral, social, or linguistic" explanations, but he wrongly thinks that the biological sciences can provide an explanation. There is no explanation.

There is a second respect in which the term "biological naturalism" is ambiguous and potentially at odds with the quadrants view. As I explained in chapter 10, it is very likely that experiential properties have causal dependencies on fundamental properties of electrons and molecular nuclei—fundamental outwardly observable properties as well as fundamental remote properties. Such properties are present in living systems, but they are not present only in living systems. They are present wherever there are electrons and molecular nuclei. If the "biological" in "biological naturalism" means that each and every property that plays a role in the generation of experience is present only in living systems, I reject this label. Searle's relentless focus on nerve cells suggests that he might intend "biological" in this strong sense. What is true, I think, is that it is only in living systems that all the properties that contribute to the generation of consciousness come together. If the "biological" in "biological naturalism" means only that, the quadrants view might qualify as a form of biological naturalism.

In sum, like "materialism" and "physicalism," the term "biological naturalism" is crucially ambiguous. The quadrants view either does or does not fall under this term, depending on how the term is understood.

14

The Quadrants and Colin McGinn: His Body-Plus Penchant

For the most part, Colin McGinn uses the word "consciousness" rather than "experience." As I understand him, he uses the words "conscious" and "consciousness" in the broad sense that is coextensive with experience: an animal is conscious and has consciousness if and only it has some features of experience. In comparing McGinn's view with the quadrants view, I use the word "consciousness" quite a bit, and always in this broad sense.

<>

Although McGinn has written sentences that suggest a body-only view, by and large he advocates a body-plus view that shares with T. H. Huxley's "mental steam" view the idea that consciousness is something emitted by the brain. While he may waver a bit on this point, he has a strong body-plus penchant that must be reckoned with.

McGinn's body-plus view differs from Huxley's in many respects. For one thing, McGinn does not share Huxley's epiphenomenalism, which is mentioned in chapter 2 and discussed more fully in chapter 22. In addition, McGinn's view is more detailed than Huxley's, and many of its details neither echo nor contradict anything that Huxley says.

One of these novel details is McGinn's idea that human beings are "cognitively closed" to certain properties that play a role in making them conscious. This means, he says, that our "concept-forming procedures ... cannot extend to a grasp of" (McGinn 1991, 3) these properties and "our minds cannot represent" (McGinn 1991, 14) them. Actually, McGinn usually suggests that there is one such property. But he sometimes uses the plural, and he gives no reason to think that there is exactly one. I therefore use the plural "properties" here. I noted in chapter 4 that most animals have little or no capacity to inwardly observe their own experience, and that the human capacity to inwardly observe experience may be limited in ways that people tend to overlook. Using McGinn's terminology, I therefore think it is possible that human beings are "cognitively closed" to some of their own experiential properties. However, it seems to me that McGinn is thinking of nonexperiential properties that human beings are "cognitively closed" to. This is how I will interpret his hypothesis throughout this discussion. Another possible source of confusion is that McGinn sometimes says that human beings are cognitively closed to these properties and sometimes that the properties are cognitively closed to human beings. To keep things simple, I will always say that human beings are cognitively closed to the properties. In a clarified form, then, this component of McGinn's view is that human beings are cognitively closed to certain nonexperiential properties—one or more of them—that play a role in producing their consciousness.

My main reason for discussing McGinn's view is to compare what I say about remote properties with what he says about properties that human beings are cognitively closed to. I cannot begin this comparison immediately, however, because I discuss remote properties in the context of a body-only view whereas McGinn discusses properties that human beings are cognitively closed to in the context of a body-plus view. Accordingly, I proceed as follows. In this chapter, I discuss problems with the body-plus aspect of McGinn's view. In light of these problems, I dismiss his body-plus view and reaffirm the conclusion of chapter 2 that a human being is an electronuclear system in the vast expanse. I then begin the next chapter by explaining how the core of what McGinn says about properties that human beings are cognitively closed to can be transferred to a body-only view. That sets the stage for the main comparison, which is between two body-only views—the quadrants view and a body-only view that resembles McGinn's body-plus view as closely as a body-only view can.

<>

McGinn uses a variety of colorful metaphors that are reminiscent of T. H. Huxley's metaphor of a nineteenth-century locomotive emitting steam:

> … consciousness seeps from organic material lumps (McGinn 1991, 119)

... there is nothing occult about how the brain secretes consciousness—no more so than the liver's secretion of bile ... (McGinn 1991, 108n25)

Cells combine and grow during gestation until the brain is mature enough to decant experiences: At first this clump of cells is without mentality, and before you know it there is consciousness throbbing away in there.... What manner of secretion is this? (McGinn 1999, 14)

When wood burns it turns into fire, and this transformation seems almost miraculous until we understand the underlying chemistry and physics. But once we do understand, we see how wood can become fire given the process of oxygenation and the energetic properties of carbon. But this is exactly the kind of understanding that eludes us when the wood of the brain ignites into the flame of consciousness. (McGinn 1999, 60)

How could the womb of the brain spill forth something so radically different from itself? (McGinn 1999, 115)

The common element in all these metaphors is one thing emitting something else of a disparate nature. What the brain emits, according to McGinn, is a kind of nonmatter stuff that coexists with the matter of the brain:

Logically, 'consciousness' is a stuff term, as 'matter' is; and I see nothing wrong, metaphysically, with recognizing that consciousness *is* a kind of stuff. (McGinn 1991, 60)

The leap from matter to mind is surely too great to be totally unmediated. (McGinn 1993, 33)

We are a mysterious amalgam of mind and matter, as are other conscious organisms. (McGinn 1999, 229)

All the indications are that evolution discovered sentience early on, conjuring it from matter as soon as sensing organisms came to be. I like to imagine that day, many millions of years ago, when the first sentient organism came along, and a brand new ingredient entered the cosmos. (McGinn 1999, 63)

In a more recent essay titled "Consciousness as a Form of Matter" (McGinn 2011), McGinn changes his terminology but not the content of his view. In this essay he uses the word "matter" more broadly to include consciousness, which he says is a "form of matter" distinct from electrons and protons.

According to the quadrants view, this is all very wrong. A human being is a sometimes-conscious electronuclear system. Consciousness is a property of the system, as are all particular features of experience. Words like "seep," "decant," "secrete," and "spill forth" are not appropriate. Consciousness is not a second sort of stuff that exists alongside electronuclear fabric. There is no leap from electronuclear fabric to something else. There is no amalgam of electronuclear fabric with something else. The evolution of sentient organisms did not add a new ingredient to the cosmos. There is only electronuclear fabric, constituting some extremely complex systems that have a wide variety of properties.

McGinn makes two fallacious arguments in support of his body-plus view.

In one of these arguments, he simply assumes the correctness of the purely outward conception of the brain that I criticized in chapter 8. He presents a version of Frank Jackson's widely discussed "knowledge argument" (Jackson 2004a and 2004b), in which Jackson imagines Mary, a supremely knowledgeable scientist who has been deprived from birth of all color experiences except black, white, and shades of gray. McGinn claims that this thought experiment shows that someone could have "complete knowledge of the brain," could know "everything about the brain," and yet know "nothing about what it is like to have conscious experiences." Therefore, he concludes, conscious experiences are something over and above the brain. But the only "knowledge of the brain" that McGinn considers in this argument is knowledge of the brain that is obtained by the outward-observation-based methods of the physical sciences. He simply assumes that complete knowledge of the brain can be obtained with these methods. This is something that there is good reason not to believe. Ironically, it is also something that McGinn himself seems to emphatically deny when he maintains that the brain has properties that human beings are cognitively closed to; I say more about this apparent inconsistency shortly.

Continuing in the same vein, McGinn discusses Thomas Nagel's claim that we have no way to know what the experiential aspects of a bat's echolocation capability are like (McGinn 1999, 18–23). "But this ignorance on our part does

185

not extend to knowledge of a bat's brain," McGinn says, again thinking only of knowledge of brains that is obtained by outward-observation-based methods of study. In saying this, he fails to consider the possibility that ignorance of the experiential aspects of a bat's echolocation capability is ignorance of certain aspects of the bat's brain. He makes no attempt to rule out this possibility, which is plausibly the case.

McGinn's purely outward conception of the brain seems to be part of a more general conception of human beings in which objects and ways of knowing are paired off in a symmetrical way. According to this conception, consciousness is open to inward observation but not to outward observation, while the brain is open to outward observation but not to inward observation. This conception surfaces in statements such as the following, which McGinn lays down as if they were uncontroversial statements of fact:

> The brain is not (as such) a potential object of introspection, and consciousness is not (as such) a potential object of perception. (McGinn 1991, 61)

> Introspective data do not (by themselves) provide us with information about the condition of our nervous system. (McGinn 1991, 73)

According to the quadrants view, these two statements are mainly false. The only truth in them is that consciousness is not a potential object of perception (outward observation). Introspection is observation of the brain and the nervous system—inward observation. This is not obviously the case, of course, but there are good reasons to think that it is the case. Consciousness (a property) is accessible only through inward observation, but the brain and the nervous system (many-propertied electronuclear objects) are accessible through both inward observation and outward observation—properties of different types being accessible in different ways. The situation does not have the symmetry that McGinn takes for granted.

I noted in passing that McGinn seems to contradict himself in basing an argument on the premise 1) that outward observation can give us "complete knowledge of the brain" while maintaining 2) that the brain has properties that we are cognitively closed to. A property that we are cognitively closed to is a property that we cannot access at all, including through outward observation. If McGinn thinks that the brain has, or even might have, even one such property, then it does not make sense for him to argue that consciousness is something over and above the brain because we cannot observe it outwardly. This inconsistency strikes me as further testimony to the strength of the grip that outwardism can have on a person. McGinn's main thesis is a challenge to the purely outward conception of the brain, yet the purely outward conception of the brain has such a grip on him that he bases a major argument on it!

McGinn's second argument for his body-plus view involves a quasi-Cartesian idea that consciousness is "nonspatial." I find McGinn's presentation of this idea so strange that I will simply quote a few relevant passages without trying to interpret them:

> The mind thus depends upon the spatial world, in the form of the brain, and it represents a spatial world, yet it steadfastly refuses to set foot in space. It just won't go there. (McGinn 1999, 111)

> The only ingredients in the pot when consciousness was cooking were particles and fields laid out in space, yet something radically non-spatial got produced. On that fine spring morning when consciousness was first laid on nature's table there was nothing around but extended matter in space, yet now a non-spatial stuff simmered and bubbled. (McGinn 2004, 100)

> The brain puts into reverse, as it were, what the big bang initiated: it erases spatial dimensions rather than creating them. It undoes the work of creating space, swallowing down matter and spitting out consciousness. (McGinn 2004, 101–102)

If the brain is spatial and consciousness is nonspatial, then it follows that consciousness is something over and above the brain. That is a logically respectable thought, but to make a persuasive argument out of it, we need a good reason to believe that consciousness is nonspatial. And this McGinn fails to deliver. He gives two reasons to believe that consciousness is nonspatial, but neither is a good one.

One reason concerns the properties of shape, size, and distance:

> Your visual experience of red or my emotion of fear has no particular shape or size. Nor does it stand in spatial

relations to other experiences. Your experience of red is not, say, next to your experience of a whistling sound, or four centimeters away from it, or behind it. There is no clear sense in the question of how great a distance separates a pair of experiences. (McGinn 1999, 109–110)

McGinn says this shows that these features of experience are not spatial. But all of this is perfectly consistent with visual experiences of red, emotions of fear, and experiences of whistling sounds being properties of an electronuclear system in the vast expanse. My weight and my temperature have no particular shape or size. It makes no sense to ask how far it is from my weight to my temperature. Should I conclude from this that my weight and my temperature are not spatial? No! My weight and my temperature are properties of my body, which is in the vast expanse. According to the quadrants view, the same is true of my consciousness and all the particular features of my experience. Nothing that McGinn says gives any reason to doubt this.

The other reason concerns competition for space:

Material objects also compete for space; that is what it is for them to be *solid*. You cannot have two things in the same place at the same time; they nudge each other aside... But it makes no sense to say that conscious states compete for space. (McGinn 1999, 110; his emphasis; see also McGinn 2004, 98)

Once again, this is perfectly consistent with features of experience being properties of an electronuclear system in the vast expanse. It makes no sense to say that my weight and my temperature compete for space or to ask whether my weight and my temperature can occupy the same place at the same time. This is because they are properties. In general, an electronuclear object can have properties galore that do not get in each other's way. And some of these properties can be features of experience.

In sum, the argument based on the idea that consciousness is nonspatial fails because McGinn gives no good reason to believe that consciousness is nonspatial. All the characteristics of consciousness that McGinn cites can easily pertain to properties of an electronuclear system in the vast expanse.

<>

McGinn uses his claim that consciousness is nonspatial to explain the fact that consciousness is not outwardly observable:

> Consciousness enables us to perceive the world, but it is not itself a perceptible thing. You cannot see someone's consciousness with your eyes or feel it with your hands, not even when the consciousness in question is your own.... This fact about consciousness is bound up with its nonspatiality. The senses respond to spatial entities by means of spatial relations. But the mind is not spatial, so the senses cannot in principle respond to it. (McGinn 1999, 113–114)

> It is not that experiences have location, shape, and dimensionality for eyes that are sharper than ours. Since they are non-spatial they are in principle unperceivable. (McGinn 2004, 95)

Believing as he does that the brain has nonexperiential properties that we are cognitively closed to, how does McGinn explain the fact that *those* properties are not perceptible? I don't know of any place where he addresses this question. According to the quadrants view, McGinn's "nonspatiality" explanation of the fact that experiential properties are not outwardly observable is erroneous. The correct explanation, in brief, is that experiential properties of electronuclear objects are properties of a type that can have no evidentiary effects on the spatial transmissions that outward observation depends on. The same explanation applies to remote properties.

I have previously touched on several other erroneous explanations of the fact that experiential properties are not outwardly observable. In chapter 7 I discussed Owen Flanagan's apparent claim that experiential properties are not outwardly observable simply because they are inwardly observable. That claim overlooks the fact that some properties can be cognitively encountered in more than one way. In chapter 8 I mentioned Fred Dretske's to-me-incomprehensible "externalist" theory that "the mind isn't in the head." In chapter 13 I discussed John Searle's idea that features of experience are not outwardly observable because they have a "subjective mode of existence" that consists in being dependent on being experienced. This supposed explanation falls flat because there is no explanatory connection between being dependent on being experienced and being closed to outward observation. Cartesian dualism, discussed in chapter 2, suggests the explanation that experiential properties are not outwardly observable because they are properties of a nonbodily "thinking thing" that has no size—either because it is not in the vast expanse or because it is a mathematical point. To these we can now add McGinn's quasi-Cartesian notion that experiential properties are not in the vast expanse despite being dependent on the body.

Many writers claim or imply, often in the name of "materialism" or "physicalism," that outward observation has no limits. That is also wrong. It is of crucial importance to appreciate that outward observation does have limits, and to explain the limits of outward observation in a plausible way. The explanation given by the quadrants view, once again, is that experiential properties (and remote properties) of electronuclear objects are properties of a type that can have no evidentiary effects on the spatial transmissions that outward observation depends upon. The full account is in chapter 7.

<>

McGinn says that he is discussing "the problem of embodiment," which concerns the "link" between consciousness and the brain. He believes this "link" is "intelligible":

> Consciousness indubitably exists, and it is connected to the brain in some intelligible way … (McGinn 1999, 5)

> Yet somehow the brain and consciousness are bound together into an intelligible whole. (McGinn 1999, 48)

Taking this belief as a premise, he makes an argument that can be summarized as follows:

> There is an intelligible link between the brain and consciousness.

> If human beings had only outwardly observable, physical-science properties and the inwardly observable experiential properties of consciousness, then there would not be an intelligible link between the brain and consciousness, because these two classes of properties have such disparate natures.

Therefore, human beings also have properties of another kind—nonexperiential properties that they are cognitively closed to—which are involved in the constitution of this intelligible link.

The conclusion of this argument sounds paradoxical. How can properties that human beings are *cognitively closed to* make the link *intelligible?* The answer is that McGinn concludes that these properties make the link intelligible, but not intelligible to human beings. That is the intriguing claim that is the heart of McGinn's view.

McGinn's discussion of this argument is tightly tied to his body-plus view, which I think is mistaken for the reasons explained in this chapter. However, as I show in the next chapter, one can make a parallel argument within the context of a body-only view.

15

The Quadrants and Colin McGinn: Intelligibility and Cognitive Closure

Someone who believes that a human being is an electronuclear system in the vast expanse could make the following argument, which preserves the logical structure of McGinn's argument:

> There are intelligible links between the body's outwardly observable physical-science properties and its inwardly observable experiential properties.

> If the body had only outwardly observable physical-science properties and inwardly observable experiential properties, there would not be any intelligible links between them, because these two classes of properties have such disparate natures.

> Therefore, the body also has properties of another kind—nonexperiential properties that human beings are cognitively closed to—which are involved in the constitution of these intelligible links.

This argument differs from McGinn's argument in two principal respects. First, it avoids the idea that consciousness is something additional to the body that the brain "secretes" or "spills forth." Second, it replaces the reference to a single "link" between consciousness and the brain with a reference to "links" between the body's experiential properties and its outwardly observable physical-science properties. The conclusion of this parallel argument also sounds paradoxical. How can properties that human beings are *cognitively closed to* make the links *intelligible*? Again, the answer is that these properties are said to make the links intelligible, but not intelligible to human beings. This is the body-only counterpart of the intriguing claim that is the heart of McGinn's view.

<>

One troublesome aspect of McGinn's work is that he uses the crucial word "intelligible" in some dubious ways. At times I get the feeling that he is so attached to this *word* that he is willing to stretch its meaning in whatever way is necessary to make the statement "There is an intelligible link between the brain and consciousness" true. I will begin by describing and setting aside three of his dubious uses of this word.

There are many adjectives that we use to indicate how certain things affect us, as opposed to the nature of the things themselves. Some common examples are "annoying," "difficult" and "enjoyable." "Intelligible" is such a word, as I understand it. Just as something is enjoyable only insofar as some being can enjoy it, so something is intelligible only insofar as some being can understand it. The following words are also of this sort:

> mysterious
> baffling
> eerie
> bizarre
> problematic
> peculiar
> dubious
> inscrutable

McGinn seems to tacitly deny this when he declares, as if stating a matter of fact, that the production of consciousness by the brain does not satisfy any of the following descriptions:

> inherently mysterious (McGinn 1991, 5)
> a case of objective eeriness (McGinn 1991, 5)
> a case of objective mystery (McGinn 1991, 5)

> inherently eerie or bizarre (McGinn 1991, 22)
> intrinsically mysterious (McGinn 1991, 31)
> intrinsically baffling (McGinn 1991, 108)
> intrinsically problematic or peculiar or dubious (McGinn 1993, 2)
> inherently inscrutable (McGinn 1993, 6)

Nothing *could satisfy* these descriptions, because they embody contradictions. If McGinn thinks that something can be eerie or baffling or peculiar in itself, and not because of its effects on some being who contemplates it, then he might think that something can be intelligible in itself even though no being can understand it. This is the first dubious suggestion that I set aside. It makes no sense to say that something is intelligible in itself. To be intelligible is to be intelligible to some being, if not some actual being then at least some feasible being.

The second dubious suggestion is that something can be intelligible to genes. McGinn flirts with this idea in the following passage:

> ... the genes work symbolically, by specifying programmes for generating organisms from the available raw materials.... So, for example, they must somehow specify the structure and functioning of the heart, and they must supply rules for generating this organ from primitive biological components. The genes are, as it were, unconscious anatomists and physiologists, equipped with the lore pertaining thereto. But what goes for the body also goes for the mind: the genes must also contain the blueprint for constructing organisms with the (biologically based) mental properties that those organisms instantiate. They must, then, represent the principles by which mental properties supervene on physical properties. They must, that is, specify instructions adequate for creating conscious states out of matter.... the genes 'know' how to construct organisms with intentionality, with personhood, with the capacity to make free choices, with rich systems of knowledge.... This requires a grip on the natural principles that constitute these attributes, as well as mastery of the trick of engineering them from living tissue. (McGinn 2004, 193–194; see also McGinn 1999, 224–228)

One problem with this passage is that it grossly exaggerates the role that genes play in the development of an organism. The main role of genes is in the manufacture of protein molecules, which account for only about 15% of the body's weight. And even here, their role is limited. Genes—DNA molecules—are templates for the linear sequences of amino acids that make up protein molecules. The way in which the strings of amino acids fold up, like a necklace stuffed into a purse, is not prefigured in the genes but is critical to the subsequent functioning of the completed protein molecules. Secondly and more fundamentally, whatever the role of genes may be in the electronuclear dance within an animal, it is fantastic to describe them as anatomists, physiologists, or engineers with blueprints that have "a grip on natural principles" and "know the answers." They are molecules, for Pete's sake. Nothing can be intelligible to genes because genes are not beings capable of thought and understanding; genes have no intellect.

The third and last dubious suggestion is a similar speculation about brains:

> If the brain is properly to perform its many regulative functions, and if this is to be done informationally, then it makes sense that it should contain some theory of how the entire machine works. Neural signals are not going to be interpretable unless embedded in some story about cerebral and bodily functioning. To put it colourfully, the brain must be a brain scientist, though not of the consciously reasoning variety: it must contain a theory of itself, useful to its functional needs.
>
> But if so, then we must allow for the possibility that it is a *better* brain scientist than those who seek to discover its nature by the use of reason. We must allow, indeed, that it might encode information about its own functioning that *could* not be represented by human reason. Perhaps it employs a level of description of its own workings that has never occurred to us, and never will. And if that were so, then what is an impenetrable mystery for human scientific reason would be no mystery for the brain's own epistemic system. For example, the mystery of how consciousness arises from the neural material might be answered by the theory the brain applies to itself in monitoring its state of consciousness …
> (McGinn 1993, 137–138; his emphasis)

Naturally, every human theorist or brain scientist has a brain that plays a major role in his or her intellectual activities, but this is the only sense in which a brain might be said to "be" a theorist or a brain scientist. Nothing can be intelligible to a dark corner of a person's brain that plays no role in the person's thinking because such brain tissue, by hypothesis, is not a being capable of thought and understanding.

For McGinn's argument or its body-only counterpart to be meaningful and interesting, the crucial word "intelligible" must mean "intelligible to some being, either actual or feasible, that is capable of thought and understanding." Granted, the being does not have to be a human being; one can certainly imagine that there are things that are not intelligible to any human being but that are intelligible to a being of another type. However, the being must be a being with intellectual capacities. DNA molecules and chunks of tissue from unthinking parts of the brain do not qualify. From here on, my discussion of the body-only counterpart of McGinn's argument employs this normal understanding of the word "intelligible."

<>

If the quadrants view is correct, the conclusion of McGinn's argument contains an important truth: human beings have nonexperiential properties that human beings are cognitively closed to. This statement is true, according to the quadrants view, because human beings have remote properties, which satisfy the specified condition. However, McGinn's conclusion as a whole is false, and the first premise of his argument is also false. According to the quadrant's view, the relation between a human being's outwardly observable physical-science properties and his or her inwardly observable experiential properties is not intelligible. It is not intelligible to human beings (a statement that McGinn agrees with) and moreover it is not intelligible to any being, either actual or feasible, that is capable of thought and understanding. Moreover, this relation is unintelligible precisely because of the involvement of remote properties in the generation of experiential properties. The properties that human beings are cognitively closed to are the reason why there is absolute unintelligibility, not the reason why there is intelligibility in some nonhuman sense. McGinn's description of the situation is upside down and backwards, even though he is right about the involvement of properties that human beings are cognitively closed to. I will elaborate on these crucial differences between

McGinn's view and the quadrants view later in the chapter. But first I am going to discuss the defects of his argument.

What is the source of McGinn's confidence in the first premise of his argument? What makes him so sure that there is an intelligible link between a human being's outwardly observable, physical-science properties ("the brain") and inwardly observable, experiential properties ("consciousness")? Many people—Searle is one example—believe there is an intelligible link because they believe there is a link that is intelligible to human beings. One frequently cited basis for this belief is humanity's proud history of successfully solving difficult problems. However, this cannot be McGinn's reason, because he does not think there is a link that is intelligible to human beings. I do not see a clear explanation in McGinn's work of the source of his confidence in his first premise, but I see two possible pieces of an explanation.

One possible source of McGinn's confidence in the first premise is an oversimplified way of thinking about correlations, which is illustrated by the following passage:

> Resolutely shunning the supernatural, I think it is undeniable that it must be in virtue of *some* natural property of the brain that organisms are conscious. There just *has* to be some explanation for how brains subserve minds.... some theory must exist which accounts for the psychophysical correlations we observe. It is implausible to take these correlations as ultimate and inexplicable facts, as simply brute. (McGinn 1991, 6; his emphasis)

There is a second illustration in the way he compares these "psychophysical correlations" with the correlation between the temperature and pressure of a fixed volume of gas that is described by Boyle's gas law. After a nice discussion of how scientists explain this temperature/pressure correlation by noting that the temperature and pressure of a gas are both manifestations of the same underlying motion of molecules, McGinn makes the following comparison:

> But if you had a mind that was prevented from forming the idea of constituent molecules and their movements, this explanation would be closed to you, and you would find yourself faced with a brute correlation. My claim is that this is essentially our predicament with respect to psychophysical

> correlations: we lack the unifying underlying theory, so we
> are deeply puzzled about the observed correlations. (McGinn
> 2004, 68–69)

What I see in both of these passages is a dichotomous contrast between theoretically explicable correlations and "brute correlations." A "brute correlation," as I understand McGinn's use of this phrase, is a correlation about which nothing illuminating can be said. The correlation exists, end of story. McGinn's tacit assumption seems to be that every correlation must either be theoretically explicable or "brute" in this sense. Rejecting the idea that the many well-established correlations between physical-science properties of the brain and experiential properties are "brute," he is therefore left with only one alternative: they are theoretically explicable. The trouble with this line of thought is that correlations that have a theoretical explanation and "brute correlations" are opposite extremes that do not exhaust the logical possibilities. Between these two extremes there is the possibility of correlations that cannot be theoretically explained yet can be made sense of to a certain extent. Following are two examples of correlations that fall in this middle category.

First, electrons have both mass and charge, and in fact they have a particular mass and a particular charge. These two properties are perfectly correlated across all electrons. They are perfectly correlated because they both belong to the indissoluble electron property package. If electrons have remote properties, then these too are correlated with the electron mass and the electron charge, and for the same reason. Describing the ground of this correlation in this way takes it out of the category of "brute correlations." Yet there may not be any theoretical explanation for it. There is no accepted theoretical explanation for the existence of tiny objects that have both mass and charge, let alone the existence of tiny objects that have precisely this mass and this charge. There is no accepted theoretical explanation for the existence of objects that have indissoluble property packages. Why is property surgery impossible? It could be that compelling theoretical explanations of all these things will be forthcoming one of these days, but it could also be that these are facts too basic to be touched by theories. In that case, the correlations between the various properties that belong to the indissoluble property packages of electrons and other fundamental particles would be neither "brute" nor theoretically explicable.

The second example is the very set of correlations that McGinn is interested in, those between physical-science properties of the brain and experiential properties. According to the quadrants view, these correlations obtain because the physical-science properties are always accompanied by remote properties that belong to the same indissoluble property packages, and the combination of all these properties jointly produces the experiential properties. Describing the ground of the correlations in this way takes them out of the category of "brute correlations." Yet the correlations are not theoretically explicable because remote properties are beyond the reach of theories. There is no possibility of explaining why certain physical-science properties of the brain are correlated with a visual sensation of red, for example, because the remote properties that are involved in the production of this sensation are not accessible. Note that the point I am making here does not depend on the quadrants view being correct. It depends only on our ability to imagine that the quadrants view might be correct, and thus to imagine that the correlations between physical-science properties of the brain and experiential properties are neither "brute" nor theoretically explicable. If we can imagine this middle category of correlations, then any argument that tacitly assumes that the two categories of "brute" correlations and theoretically explicable correlations are logically exhaustive is unsound.

The second possible source of McGinn's confidence in his intelligibility premise is his habit of imagining a supernatural omniscient being. He does not believe that such a being exists; he states "I do not believe in the supernatural in any form" (McGinn 1999, 84). Yet he invokes various conceptions of this nonexistent omniscient being with astonishing frequency. Following are a few of the many references to an all-knowing God in McGinn's writing:

> God, knowing the details of the hidden structure, can see quite plainly that there are no immaterial substances and the like, and He fully appreciates the nature of the necessities that link consciousness to the body. (McGinn 1991, 107–108)

> God, we know, feels no temptation to subscribe to a Cartesian ontology; His knowledge goes too deep for that. We should try to follow his example, despite the poverty of our understanding. (McGinn 1991, 108)

> God obviously does not apprehend the world by directing his senses toward it—seeing it, hearing it, smelling it, etc. Nor

> does he infer our mental states from our behavior. There
> is no process of divine theory construction, confirmation,
> induction, inference to the best explanation, and so on. God
> doesn't need to employ the scientific method. Somehow, we
> are to understand, God's omniscience takes in everything, in
> an instant, and with extraordinary clarity and directness; he
> just "intuits" the totality of reality. (McGinn 2004, 23)

> What I am suggesting, basically, is the existence of (humanly)
> unknowable conceptual connections between mind and
> brain. The implicated concepts are not ours, obviously....
> Heuristically, we can think of them as the contents of God's
> thoughts.... God can see right into the conceptual necessities
> whose existence we can only gesture at. (McGinn 2004, 51)

A habit of imagining an omniscient being is a habit of imagining that *everything* is intelligible—to that being. It is a small step, psychologically speaking, from *imagining* that everything is intelligible to *believing* that everything is intelligible, and thus believing that the links between the body's experiential properties and its outwardly observable physical-science properties are intelligible. To my knowledge, McGinn never claims that absolutely everything that exists is fully intelligible. But neither does he ever deny this. Moreover, the following appeal to the nonsensical idea of "the point of view of Nature" can be seen as an attempt to legitimate an omniscient point of view that does not belong to an actual being:

> It is comforting to reflect that from God's point of view, *i.e.*
> the point of view of Nature, there is no inherent mystery
> about consciousness at all. (McGinn 1991, 43)

It seems possible, then, that McGinn's confidence in the first premise of his argument rests in part on a vague notion of total intelligibility that is nourished by his habit of imagining an omniscient being.

McGinn is not alone in thinking that it can be illuminating to imagine an omniscient being's perspective on the world even if there is no such being. It is in fact fairly common for writers who do not believe in God to invoke the idea of divine omniscience in order to make some point or other. I think this practice spreads confusion and provides no redeeming benefits. I am aware of no good reason to believe in the feasibility of thinking and understanding beings that are not electronuclear systems. All the thinking and understanding beings that we have any evidence for are electronuclear systems. If there are extraterrestrials that can think and understand, they are electronuclear systems. If human beings someday build a computer that can think and understand, it will be an electronuclear system. In sum, there is a strong case that thinking and understanding are exclusively capabilities of certain suitably constituted electronuclear systems. Thinking and understanding cannot take place outside an electronuclear system any more than photosynthesis and muscle contraction can. This is not obvious, but there is good reason to believe it. And if it is true, the very act of imagining a state of understanding that is not a state of an electronuclear system involves a fantasy that sheds no light on the real world. Descriptions of how an omniscient being views the world should be left to those who believe that such a being exists.

<>

I have said that nothing is intrinsically intelligible, because intelligibility consists in a relation to a being that has a capacity to understand. I have said that nothing is intelligible to unthinking objects such as DNA molecules, because they have no capacity to understand. I have said that nothing is intelligible to a supernatural omniscient being, because there is no such being. The only remaining possibility, I believe, is intelligibility to an electronuclear thinker, either actual or feasible. The links between the body's experiential properties and its outwardly observable physical-science properties are intelligible if and only if they are intelligible to some actual or feasible electronuclear thinker.

McGinn makes a case for various possibilities in this category based on the fact that the animals we are acquainted with differ markedly in their "concept-forming capacities" (McGinn 1991, 10, 29, 42, 104, 119, 120). It is easy to imagine that there are or could be thinkers with concept-forming capacities that human beings lack, just as human beings have concept-forming capacities that dogs and turtles lack. Such thinkers might have all the concept-forming capacities that human beings have, plus more, or they might have some concept-forming capacities that human beings lack while lacking some that human beings have. Either way, they might be able to think thoughts that human beings cannot think, including thoughts that enable them to understand the links between the body's experiential properties and its physical-science properties.

McGinn speculates that "beings from another galaxy" (McGinn 1999, 213) might have this ability, or that it might be possible to produce this ability through selective breeding (McGinn 1999, 219) or genetic reengineering (McGinn 1999, 220) of today's human beings. The different ways in which a superhuman electronuclear thinker might come into existence—evolution on another planet, selective breeding, genetic reengineering, or even a high-tech manufacturing process—are incidental to McGinn's argument. The key point is that we have no reason to believe that human beings stand at the summit of nature's capacity to produce concept-forming capacities. In all likelihood, electronuclear fabric can be fashioned into thinkers of other kinds, and among these there could be thinkers who can understand the links between the body's experiential properties and its physical-science properties.

I agree with McGinn, as I think everyone must, that there is no reason to think that the human beings of today represent the ultimate in electronuclear thinking capacity. Superhuman thinkers might well be feasible. But note that this argument is utterly generic. Its conclusion concerns thinking in general, not thinking about the mind-body problem in particular. It might be that superhuman mathematicians, musicians, and chess players are all feasible, but that superhuman solvers of the mind-body problem are not feasible. The human perspective on the links between a body's experiential properties and its physical-science properties might be the very same perspective that any highly intelligent electronuclear thinker will have. According to the quadrants view, this is precisely the case. There is a limit to the ability of any electronuclear system to understand these links, and human beings have reached it.

There is a footnote in one of McGinn's books that nicely highlights the difference between his view and the quadrants view concerning this important point. McGinn states that "it is something about the tracks of our thought that prevents us from achieving a science that relates consciousness to its physical basis: *the enemy lies within the gates*" (McGinn 1991, 19; my emphasis). To this statement he attaches a footnote that begins as follows:

> I get this phrase from Fodor, *The Modularity of Mind*, p. 121. The intended contrast is with kinds of cognitive closure that stem from exogenous factors—as, say, in astronomy. (McGinn 1991, 19n)

The exogenous factors exemplified by astronomy are those that make it difficult or impossible for earthlings to observe certain distant objects. Here are three such exogenous factors:

- Some objects are so far away that the spatial transmissions they emit have not reached the earth yet. In some cases the spatial transmissions will never reach the earth due to the ongoing expansion of the vast expanse.
- Some distant objects do not emit spatial transmissions in sufficient strength to enable us to observe them.
- Some distant objects are eclipsed by closer objects that block the spatial transmissions of the more distant objects before they can reach us.

A key claim of the quadrants view is that our cognitive encounters with all electronuclear objects, regardless of their distance from us, are limited by an exogenous factor that has some resemblance to these three: all the electronuclear objects that we do observe, either with our unaided senses or with the help of instrumentation, have *properties* that we cannot observe because they are properties of a sort that can have no evidentiary effects on the spatial transmissions. We observe electronuclear objects by observing their outwardly observable properties, which are not all their properties. The exogenous factors that McGinn alludes to are due to the absence of any evidence-carrying connection from a given object to an observer: no spatial transmissions complete the trip to an observer's sense organs. The exogenous factor relevant to the quadrants view is due to the absence of any evidence-carrying connection from certain properties of an electronuclear object—its remote properties—to the spatial transmissions that the object emits. The object emits spatial transmissions, which reach observers and enable them to observe the object, but only the object's outward profile. In McGinn's terminology, observers are "cognitively closed" to the observed object's remote properties due to this exogenous factor.

This exogenous factor applies uniformly to all actual or feasible electronuclear beings, precisely because it is exogenous. Moreover, this exogenous factor suffices to make the links between any being's experiential properties and its physical-science properties unintelligible to that being, because the generation of the experiential properties depends in part on properties of electronuclear fabric that can have no evidentiary effects on the spatial transmissions. In consequence, no feasible electronuclear being can understand the links between its experiential properties and its physical-science properties. An electronuclear being might well have superhuman powers of various sorts "within its gates," but there is no power that can extract from the spatial transmissions that strike its sense organs what is not in those transmissions.

McGinn writes:

> What I want to suggest is that the nature of the psychophysical connection has a full and non-mysterious explanation in a certain science, but that this science is inaccessible to us as a matter of principle. (McGinn 1991, 17)

> If we imagine creatures whose cognitive structure allows them to incorporate the mystery-resolving senses into their thoughts, then we can say that for these creatures there is no aura of impenetrable mystery surrounding the psychophysical link. For them, the connexion is as unmysterious as any other natural nexus, a matter of plain science…. In the case of the mind-body problem, the mystery exists only for creatures whose cognitive slant biases them away from the concepts that are needed to make the question into a mere scientific problem. (McGinn 1993, 40–41)

According to the quadrants view, the way in which experiential properties are generated does not have any explanation, scientific or otherwise, because any electronuclear being that can construct explanations will be cognitively cut off from the remote properties of electronuclear fabric for the same exogenous reason. There can be no concepts of particular remote properties because there is a barrier that cannot be breached, namely the mismatch between the plenary nature of electronuclear objects and the nature of the spatial transmissions. These remote properties, causally crucial in the generation of experience, can have no cognitive representation. They lie beyond all feasible thought. They are universally unintelligible.

McGinn writes:

> … it is just a matter of bad cognitive *luck* that we cannot solve the mind-body problem; our minds happen not to have been engineered that way. (McGinn 1999, 214; his emphasis)

According to the quadrants view, the obstacle to understanding is not how human beings are "engineered" but how the universe is "engineered." This universe, containing objects made of this electronuclear fabric, traversed by these spatial transmissions, cannot produce a thinker who can make cognitive contact with all the properties of electronuclear fabric that contribute to the generation of experiential properties. An electronuclear thinker can only extract from the spatial transmissions what is in them, regardless of how that thinker is "engineered."

McGinn contrasts "our own incurable cognitive poverty" (McGinn 1991, 43) with the problem-solving cognitive wealth of an imagined being. According to the quadrants view, the "incurable cognitive poverty" belongs to the universe, which cannot produce a cognitive system that is capable of understanding the remote properties of electronuclear objects. What can occur in the universe is limited in various ways, and this is one of them. Nothing can travel faster than light. Total entropy cannot decrease. No electronuclear being can discover through outward observation all the properties of electronuclear fabric that play a role in generating experience. I should add that there is an element of arbitrariness in describing this situation by saying that the universe is cognitively poor. One can make the same point by saying that the universe is *propertywise rich* beyond what any thinker it can produce can understand. Or by saying, more neutrally, that the universe's potential to produce cognitive encounters with properties does not extend to all the properties that it has.

McGinn introduces the term "transcendental naturalism" to highlight his idea that there are respects in which nature transcends the ability of human beings to understand it. This is a nice phrase and a valid thought, but McGinn mistakes the locus and nature of the transcendence that is relevant to the mind-body problem. There may well be some aspects of nature that are open to "forms of understanding" that "are not humanly accessible" (McGinn 1993, 18). But there are also aspects of nature that are open to no form of understanding, because nature transcends its own ability to produce beings that can understand it. It is this global or universal transcendence that cloaks the links between experiential properties and the physical-science properties on which they depend.

In *Mind and Cosmos*, Thomas Nagel writes that he "assumes the fundamental intelligibility of the universe" (Nagel 2012, 47). Here are key excerpts from a passage in which he elaborates this idea:

> This assumption is a form of the principle of sufficient reason—that everything about the world can at some level be understood ...

205

> The view that rational intelligibility is at the root of the natural order makes me, in a broad sense, an idealist— ...an objective idealist in the tradition of Plato ...
>
> The intelligibility of the world is no accident. Mind, in this view, is doubly related to the natural order. Nature is such as to give rise to conscious beings with minds; and it is such as to be comprehensible to such beings. Ultimately, therefore, such beings should be comprehensible to themselves. (Nagel 2012, 17)

Except for the statement "Nature is such as to give rise to conscious beings with minds," there is no reason to believe any of this. If there is no supernatural omniscient being—and Nagel does not think there is—then the degree to which the world is intelligible depends on the world's ability to produce beings that can understand it. The existence of human beings shows that the world can produce beings that can understand it *to some extent,* but can the world produce beings that can understand it fully? According to the quadrants view, it cannot, because electronuclear fabric has properties that no feasible electronuclear being can discover. Nagel gives no reason to think otherwise. In what sounds a bit like a reason, he says "Science is driven by the assumption that the world is intelligible" (Nagel 2012, 16). This statement has a nice ring to it, but it does not stand up to scrutiny. It is obvious—not an assumption—that the world is *somewhat* intelligible. This fact gives scientists all the support they need. "Let's learn what we can" is a good and sufficient motto for the scientific community. It is unnecessary—and unscientific!—to assume that the world is *fully* intelligible.

In a departure from his main position, McGinn writes:

> How strong is the thesis I am urging? Let me distinguish *absolute* from *relative* claims of cognitive closure. A problem is absolutely cognitively closed if no possible mind could resolve it; a problem is relatively closed if minds of some sorts can in principle solve it while minds of other sorts cannot.... It certainly seems to me to be at least an open question whether the problem is absolutely insoluble; I would not be surprised if it were. (McGinn 1991, 15–16; his emphasis)

This passage is at odds with McGinn's intelligibility argument, assuming that intelligibility means intelligibility to some possible mind. According to the quadrants view, this passage is also closer to the truth than McGinn's intelligibility argument, because there is absolute cognitive closure regarding the way in which experiential properties are generated. Although McGinn was not thinking of the quadrants view when he wrote this passage, it appears that when he wrote this passage he was open to the possibility that this implication of the quadrants view might be correct.

Some writers discuss "possible worlds." Others speculate about "parallel universes." Imagine conscious beings that live in a world that contains no electrons, protons, neutrons or electromagnetic radiation. These beings are made of a kind of stuff that does not exist in our vast expanse, and they have brains and sense organs that work in ways utterly unlike our own. Might they be able to understand how their experiential properties are generated? I have nothing to say about such wide-open fantasies. The quadrants view and my arguments for it pertain "only" to the vast expanse that we are in.

Note that my discussion of electronuclear thinkers and my earlier discussion of supernatural omniscience are logically independent of each other. It would be logically consistent to agree with me that the quadrants view is true of the vast expanse and its inhabitants while believing that there is a supernatural omniscient being. However, the supernatural omniscient being would have to be of the deistic sort, uninvolved in human affairs. He could not, for example, preside over an afterlife where human beings live on after their hearts stop beating, because immortality of that sort is incompatible with the quadrants view.

<>

An important part of the quadrants view is the hypothesis that experiential properties are joint products of outwardly observable properties and remote properties. I have scoured McGinn's writings for a clear statement of a causal hypothesis to compare with this one, but I have not found any. The reason, I believe, is that the "link" that McGinn dreams of is a conceptual, thinkable link that would make the relation between "consciousness" and "the brain" intelligible. He believes that there are associated causal relations, but the causal configuration is of secondary importance for him.

In a couple of places, he seems to suggest a two-step causal sequence in which physical-science properties of the brain produce the properties that human beings are cognitively closed to, which in turn produce consciousness. Such a configuration is at odds with the quadrants view, according to which experiential properties are jointly produced by outwardly observable properties and remote properties. The causally relevant properties that human beings are cognitively closed to are not causally *between* physical-science properties and experiential properties, like the output of the first step of a two-step manufacturing process. Rather, they are causally *alongside* physical-science properties, like one of two men who together lift a heavy object. As I noted in chapter 10, the Greek letter π makes a good icon for this causal configuration. Considering McGinn's writing as a whole, however, it is not clear to me that he is committed to the idea of a two-step causal sequence with the properties that human beings are cognitively closed to causally in the middle. What he argues for is the existence of a conceptual link that makes the causal situation intelligible, whatever that situation might be. Accordingly, I would say that the π configuration of the quadrants view contrasts with McGinn's openness concerning the causal situation and his lesser degree of attention to it. McGinn's main concern is intelligibility.

<>

In his 1991 book *The Science of the Mind*, Owen Flanagan gave McGinn's view the name "mysterianism." This term caught on, and since then many writers have referred to McGinn as a mysterian. The similarities between McGinn's view and the quadrants view are such that I can imagine someone applying this term to the quadrants view as well. This prospect prompts the following three remarks.

First, there is a sense in which the quadrants view is more mysterian than McGinn's view. McGinn holds that the link between "consciousness" and "the brain" is not intelligible to human beings, but is intelligible in some nonhuman way. According to the quadrants view, the links between the body's experiential properties and its physical-science properties are not intelligible in any way. No feasible electronuclear being in the vast expanse can understand them. This makes the relation between experiential properties and physical-science properties a more thoroughgoing mystery than it is according to McGinn.

Second, there is also a sense in which the quadrants view is less mysterian than McGinn's view. McGinn never explains why the "link" should be unintelligible to human beings but intelligible to beings of another kind. He merely makes the generic point that this could be the case because creatures differ in their "concept-forming capacities." His view thus includes a second mystery that concerns the reason for the existence of the main mystery. The quadrants view, on the other hand, includes no second mystery because it explains why the main mystery exists. The links between a body's experiential properties and its physical-science properties are not intelligible to any feasible electronuclear being in the vast expanse because they consist in part of remote properties that have no evidentiary effects on the spatial transmissions on which all outward observation depends. The quadrants view affirms the existence of one deep mystery, but it explains why this mystery exists.

Third and finally, I don't much like the term "mysterianism" because of its negative connotations. Disparagement was part of Flanagan's intent. His book includes condescending remarks such as "the new mysterians are a postmodern group, naturalists with a kinky twist" and "the doctrine is reactionary and extremely mischievous" (Flanagan 1991, 313–314). For me, the term brings to mind movie scenes of mystic trances and mist-shrouded mansions at midnight. There is an insinuation that any adherent of a "mysterian" view must have a screw loose. It is true that some writers, including McGinn himself, have used the term without derogatory intent, but it retains its negative flavor nevertheless. I therefore advise against calling the quadrants view a form of mysterianism. The quadrants view affirms a deep mystery on the basis of what I believe are strong considerations. If the considerations are strong and the mystery is real, the view does not deserve a derogatory label.

Some writers have called McGinn a "defeatist" concerning the problem of explaining how experience is generated. This is another pejorative term that I can imagine a critic of the quadrants view applying to me. In anticipation, I reply as follows. Whether it is appropriate to call someone a defeatist depends on the nature of their reason for not trying. People who do not try to find the largest prime number or to build a perpetual motion machine do not deserve to be called defeatists. I think McGinn leaves himself open to the charge of being a defeatist because his position is fallaciously defended and only partly right, as explained in this chapter. However, I think I have given a good reason for not trying to explain how experience is generated, or at least for approaching that task with minimal expectations. I have explained why it is reasonable to believe that electronuclear fabric has remote properties—nonexperiential properties that can have no evidentiary effects on the spatial transmissions. And I have explained why it is reasonable to believe that some of these remote properties are involved in the generation of experiential properties. If both of these things are true, then the goal of explaining how experience is generated is unachievable, not only for human beings but for all electronuclear beings in the vast expanse. These considerations are not conclusive, but I think they are strong enough to give serious pause to anyone who hopes to explain how experience is generated.

This is by no means a recommendation to stop studying brains or to stop speculating about how brains work. There is unquestionably much more of practical value as well as of intellectual interest that human beings can learn about brains. My point is only that these studies will never yield an explanation of how experience is generated because they cannot include the participating remote properties.

16

The Quadrants and Bertrand Russell: Events

This is the first of two chapters on Bertrand Russell. It focuses on his speculations concerning electrons, protons, and brains. The chapter that follows focuses on what Russell says about outward observation—or, to use his terminology, making inferences from perception. Russell's views went through various changes in the course of a long life. The views that I discuss here appear in *The Analysis of Matter* and some of his other books, but they are not consistent with everything he wrote.

<>

In the first chapter of *The Analysis of Matter*, Russell introduces one of the themes of that book in the following way:

> We shall find, if I am not mistaken, that the objects which are mathematically primitive in physics, such as electrons, protons, and points in space-time, are all logically complex structures composed of entities which are metaphysically more primitive, which may be conveniently called "events." It is a matter for mathematical logic to show how to construct, out of these, the objects required by the mathematical physicist. (Russell 2007, 10)

This passage does not mention neutrons because it was published in 1927, five years before neutrons were discovered. One can safely assume that Russell would have said the same about neutrons if he had heard of them. I am not concerned here with Russell's discussion of "points in space-time." My topic is his distinctive conception of electrons, protons, and by implication neutrons—the constituents of electronuclear fabric.

As Russell notes in this introductory paragraph, he uses the word "event" for the "entities which are metaphysically more primitive" than electrons and protons. I consider this a poor choice of terminology because there are huge differences between the everyday use of the word "event" and Russell's use of it to designate these hypothesized primitive entities. One cannot read the word "event" without reacting to it in an everyday way, but in reading about Russell's primitive "events" that reaction is not appropriate. Moreover, it seems to me that Russell sometimes uses the word "event" in its everyday sense, which is hardly surprising. The consequence is a troublesome lack of clarity in many of Russell's uses of this important word.

To avoid the confusion that Russell's unusual use of the word "event" can breed, I am going to introduce a different term for the primitive entities in question. I am going to call them *ephemerons*. This term has several advantages. First, it highlights a salient characteristic of these entities—that they have an extremely brief existence. Second, it ends with "on," which signals the close relation of these entities to electrons and protons. Third and finally, it is close in spirit to two terms that Russell himself used in an earlier essay for what seems to be the same idea—"momentary particulars" and "evanescent particulars" (Russell 1959, 137). In my opinion, Russell would have done better to make one of these phrases his standard term for the "metaphysically more primitive" entities, instead of repurposing the word "event."

The ephemerons of Russell's theory differ from events in the everyday sense in the following three ways.

First, an ephemeron is a spatially and temporally definite building block:

> An "event", as I understand it, is something having a small finite duration and a small finite extension in space; or rather, in view of the theory of relativity, it is something occupying a small finite amount of space-time. (Russell 1995b, 222)

> I propose to regard events as occupying regions of space-time which, in some possible metric, are spheres so far as their space-dimensions are concerned, and between a certain

maximum and a certain minimum so far as their time-dimension is concerned. (Russell 2007, 299)

In Russell's theory, ephemerons have the same kind of definite and fundamental status that Lego-like atoms have for Lucretius. Events in the everyday sense, on the other hand, are not definite things with well-defined boundaries. People typically use the word "event" to direct attention to aspects of the world's ongoing totality that interest them. For example, a "sporting event" includes certain activities of the athletes and fans, but does it include everybody's travel to and from the stadium or their daydreams and digestive chemistry during the game? There is no way to answer such questions and no need to, because "events" in the everyday sense are inventions that reflect the interest that human beings take in things. In composing the phrase "When in the course of human events" Thomas Jefferson was writing idiomatically about the seamless flow of human history; he was not discussing a countable set of definite items. A search for the boundaries in space or time of a particular "human event" would be misconceived, futile, and pointless.

Second, the hallmark of an event in the everyday sense is internal activity, whereas an ephemeron is absolutely static between its coming-to-be and its ceasing-to-be. The hallmark of an ephemeron is a very brief existence—generally a second or less—during which time it exists in a fixed state.

Third, an event in the everyday sense involves participating electronuclear objects. A sporting event involves players, fans, and usually a ball. A human event in Thomas Jefferson's sense involves human beings, of course, and perhaps some man-made artifacts such as buildings, machines, or tools. But an ephemeron does not involve participating electronuclear objects, because it is "metaphysically more primitive" than any electronuclear object. According to Russell's theory, the things that I call electronuclear objects are dense swarms of ephemerons in four-dimensional space-time.

<>

Russell says that electrons and protons are "logically complex structures" of ephemerons, but he does not make it clear what type of structure he thinks is involved. He describes structures of three different types, none of them in detail, and without ever acknowledging that he has described more than one type of structure. It's as if the three types of structure were competing for his allegiance and the question never got resolved.

In *The Analysis of Matter,* the type of structure that he mentions most often is a temporal series, or string, of ephemerons, each connected to the next in a certain way:

> We have a series of events connected together by causal laws; these may be taken to *be* the electron ... (Russell 2007, 244–245; his emphasis)

> Strings of events exist which are connected with each other according to the laws of motion; one such string is called one piece of matter, and the transition from one event in the string to another is called a motion. (Russell 2007, 246)

Any motion or other change that an electron or proton undergoes is due to a rapid-fire succession of individually static ephemerons that differ from one another in certain ways:

> The world ... consists of a number of events, each involving no change within itself, but each connected with earlier and later events by quantum or other laws ... (Russell 2007, 372)

> ... motion from one point to another consists in the cessation of certain events and the coming into existence of others ... (Russell 2007, 379)

> ... motion loses its fundamental character, being replaced by successions of events belonging to the biographies of bits of matter. (Russell 2007, 380)

In an analogy that I find very helpful, he compares the rapid-fire succession of individually static ephemerons to the rapid-fire succession of individually static movie frames projected onto a movie screen:

> A piece of matter, which we took to be a single persistent entity, is really a string of entities, like the apparently persistent objects in a cinema. (Russell 2004a, 61)

In the cinema, we seem to see a man falling off a skyscraper, catching hold of the telegraph wires, and reaching the ground none the worse. We know that, in fact, there are a number of different photographs, and the appearance of a single "thing" moving is deceptive. In this respect, the real world resembles the cinema. (Russell 1995b, 94)

In sum, what people think of as long-lived electrons and protons are in reality causally connected single-file sequences of extremely short-lived ephemerons.

This is all fairly clear, but I have so far presented descriptions of only one type of structure. The following passages describe what seems to be a second type of structure:

... an electron at an instant is a grouping of events; the question is: what sort of group is it? Obviously it includes all the events that happen where the electron is. (Russell 2007, 320–321)

Electrons and protons are not events, according to my theory; they are series of groups of events. (Russell 1963, 685)

Perhaps "a series of events" and "a series of groups of events" are two ways of describing the same structure. Russell occasionally alludes to a distinction between "elementary events" (2007, 292) or "minimal events" (1995b, 223) and more complex "events" that somehow contain many elementary events. So perhaps "a series of events" can be understood as a series of complexes, each of which is composed of several ephemerons. However, if this way of reconciling the two descriptions is correct, then Russell's writing is highly misleading. He never gives any indication that the "events" in a "series of events" have any internal structure, and he never explains what determines whether two simultaneously existing ephemerons are members of the same complex or not. It is at least as plausible, in my opinion, that Russell wrote these different descriptions on different occasions because he was imagining genuinely different types of structure.

Yet a third type of structure is described in the following passages:

It is chiefly through ideas derived from sight that physicists have been led to the modern conception of the atom as a centre

215

from which radiations travel. We do not know what happens
in the centre. The idea that there is a little hard lump there,
which is the electron or proton, is an illegitimate intrusion
of common-sense notions derived from touch. For aught we
know, the atom may consist entirely of the radiations which
come out of it. (Russell 1995b, 124–125)

Heisenberg regards a piece of matter as a centre from which
radiations travel outward; the radiations are supposed to
really occur, but the matter at their centre is reduced to a
mere mathematical fiction. (Russell 1995b, 224)

Light comes out of an atom in certain circumstances, and
the modern physicist suggests that the atom is the light that
comes out of it. Consider for a moment what happens when,
as we say, we "see" an electric light or a gas flame. Certain
waves travel over the region between the light and the eye,
and they then set up processes in the eye and the optic nerve
which end in a sensation. There is no reason to suppose that
the waves come out of a "thing"; so long as the waves reach the
eye we shall have the sensation, quite regardless of whether
they come from a "thing" or not. And so matter has become
nothing but a system of waves travelling about—not waves
"in" a sea, or "in" air or "in" the ether, but just waves. (Russell
1996, 30)

The suggestion here seems to be that electrons and protons have no reality
at all apart from certain patterns of electromagnetic radiation that people
wrongly assume to be radiating *from* something. If this were the case, only the
ephemerons that constitute electromagnetic radiation would exist. There would
be no ephemerons of distinctively electron-constituting or proton-constituting
types. I question Russell's attribution of this view to Heisenberg, but Russell
himself obviously had some affection for it when he wrote these passages.

<>

These sketchy descriptions of structure prompt various questions about
ephemerons that Russell does not discuss. One thing that bothers me is the idea

that ephemerons come into existence and go out of existence instantaneously. Does it make sense for such stark changes to occur without any passage of time? I also wonder what Russell would have said if asked whether anything causes an ephemeron to go out of existence. He says that each ephemeron causes its successor to come into existence, but once an ephemeron is in existence why doesn't it remain in existence indefinitely instead of promptly "going poof"? One can wish that Russell had gone into questions such as these, but one has to take his presentation as it stands.

Here is a more important question. Do electrons and protons really have a structure of one of these three types? Do any of Russell's descriptions of the ephemeron structure of electrons and protons have any truth to them? As far as I can tell, Russell does not give a good reason to believe that they do.

I'll start with the third type of structure, according to which electrons and protons are nothing more than patterns in the ubiquitous flux of electromagnetic radiation. Note that the first passage describing this type of structure presents a false choice between radiation from a "little hard lump" and radiation from nothing. The usual conception of electrons and protons is neither of these. The case against the radiation-from-nothing theory is overwhelming. If electromagnetic radiation is not reflected, refracted, and diffracted by something other than itself, then the patterns of electromagnetic radiation that we call reflection, refraction, and diffraction have no explanation. If there are not electronuclear molecules in addition to electromagnetic radiation, then we must give up our current understanding of all sensory stimuli except visual stimuli. If there are not electronuclear molecules in addition to electromagnetic radiation, then we have no explanation for the weight of everyday objects, since electromagnetic radiation has no weight. Given considerations such as these, I think one can confidently reject the idea that electrons and protons are merely patterns in the flux of electromagnetic radiation.

The first and second types of structure can be conveniently assessed together. Russell presents two arguments for such structures that are based on relativity theory. These arguments are not easy to follow, but insofar as they are understandable they are not sound.

The following three passages present variations of one argument:

> We can no longer speak of a body at a given time, but must speak simply of an event. Between two events there is, quite independently of any observer, a certain relation called the "interval" between them. This interval will be differently

analysed by different observers into a spatial and a temporal component, but this analysis has no objective validity. The interval is an objective physical fact, but its spatial and temporal elements are not.

It is obvious that our old comfortable notion of "solid matter" cannot survive. A piece of matter is nothing but a series of events obeying certain laws. (Russell 2004a, 60)

The substitution of space-time for space and time has made the category of substance less applicable than formerly, since the essence of substance was persistence through time, and there is now no one cosmic time. The result of this is to turn the physical world into a four-dimensional continuum of events, instead of a series of three-dimensional states of a world composed of persistent bits of matter. (Russell 1995b, 236)

Einstein ... showed that between two events there is a relation, which we may call "interval," which can be divided in many different ways into what we should regard as a spatial distance and what we should regard as a lapse of time. All these different ways are equally legitimate; there is not one way which is more "right" than the others. The choice between them is a matter of pure convention, like the choice between the metric system and the system of feet and inches.

It follows from this that the fundamental manifold of physics cannot consist of persistent particles in motion, but must consist of a four-dimensional manifold of "events." (Russell 1948, 19)

Russell's thought here seems to be that we must abandon the belief that there are persisting objects because there is no uniquely correct way to measure how long something persists. This reasoning is doubly flawed. First, as long as every observer measures some duration or other, it makes perfectly good sense to say that the observed object persists. The fact (if it is a fact) that observers moving at different velocities relative to the observed object obtain different numerical values for the duration is not a threat to persistence as such. Second, if persistence really were a problem, Russell's theory would not be the solution, because ephemerons persist. They do not persist very long by human standards, but that is irrelevant. Russell does not make a quantitative argument that there is some upper limit on how long something can persist; he makes a qualitative argument that persistence as such is impossible. This same confusion surfaces in a different context in the following passage:

> It is because I regard single observations as supplying our factual premisses that I cannot admit, in the statement of such premisses, the notion of "thing," which involves some degree of persistence, and can, therefore, only be derived from a plurality of observations. (Russell 1940, 396)

This reasoning assumes that a single observation cannot involve "some degree of persistence," which is false. Making an observation takes time. Even if one were to stipulate that a new observation begins whenever the observer blinks, there would still be "some degree of persistence" from one blink to the next. Throughout his writing, Russell seems to confuse brief persistence with no persistence; this is a fundamental mistake.

The following passage seems to argue from relativity theory to the ephemeron theory of electrons and protons in a different way:

> In the old view, a piece of matter was something which survived all through time, while never being at more than one place at a given time. This way of looking at things is obviously connected with the complete separation of space and time in which people formerly believed. When we substitute space-time for space and time, we shall naturally expect to derive the physical world from constituents which are as limited in time as in space. Such constituents are what we call 'events.' An event does not persist and move, like the traditional piece of matter; it merely exists for its little moment and then ceases. A piece of matter will thus be resolved into a series of events. Just as, in the old view, an extended body was composed of a number of particles, so, now, each particle, being extended in time, must be regarded as composed of what we may call 'event-particles.' The whole series of these events makes up the whole history of the particle, and the particle is regarded as being its history, not some metaphysical entity to which

219

the events happen. This view is rendered necessary by the fact that relativity compels us to place time and space more on a level than they were in the older physics. (Russell 1997, 142–143)

Russell's thought here seems to be that anything that is narrowly bounded spatially must also be narrowly bounded temporally, because of the way that relativity theory fuses space and time into space-time. This is very loose thinking. Relativity theory does not abolish all differences between space and time. On the contrary, it affirms that the spatial and temporal dimensions of space-time are different in significant ways. One of the differences could easily be that things can be spatially very small without being temporally short-lived. Electrons, protons, and neutrons might be examples.

Here and there Russell makes statements that seem to be at odds with his arguments from relativity theory. Here are two examples:

An electron *may* be a "thing," but it is absolutely impossible to obtain any evidence for or against this possibility ... (Russell 2007, 247; his emphasis)

But in the case of the electron, we think of it as a single persistent entity. There *may* be such an entity, but we can have no evidence that there is. (Russell 1995b, 93; his emphasis)

Russell believes that there is evidence for relativity theory. If he thinks that his ephemeron theory follows from relativity theory, as his arguments seem to say, then he should regard the evidence for relativity theory as evidence for his ephemeron theory. Yet these passages seem to say that there is no evidence for his ephemeron theory. Perhaps his point here is that there is no *direct* evidence for his ephemeron theory—no way to detect individual ephemerons or to measure their shapes, sizes, and durations, for example. Even so, it is hard to see how Russell can be serious about his anti-persistence arguments from relativity theory and yet remain so open to the possibility that the conclusion of those arguments is wrong.

Here is another riddle concerning Russell's view of the relation between his ephemeron theory and scientific evidence. In the passages just quoted, he seems to say not only that there is no currently existing evidence for the ephemeron theory, but that there can never be any evidence for it. Moreover, nowhere in his writing does he raise the possibility that physicists of the future, equipped with more sensitive instrumentation, might devise ways to detect ephemerons and to measure their shapes, sizes, and durations. He seems to think that close scrutiny of electrons and protons in the laboratory is simply not relevant to his ephemeron theory. But why should that be? The properties that he attributes to ephemerons, including beginnings and endings, shapes, sizes, and durations, are properties that physicists routinely detect and measure for objects of many kinds. If the career of an electron or a proton really has the staccato, movie-like structure of a series of individually static ephemerons, one might expect advances in observational technology to make that structure manifest. Just as someone who becomes curious about motion picture technology can observe the staccato structure of a reel of film, physicists who study what seem to be persisting electrons and protons should eventually be able to observe the staccato structure of a sequence of ephemerons—if that is what electrons and protons really are. Why then doesn't Russell discuss the possibility of technological advances that could produce observational confirmation of the ephemeron structure of electrons and protons? I have no idea how to answer this question; Russell's view of the relation between his ephemeron theory and laboratory physics is a complete mystery to me. This is the first of several examples of conflict between Russell's theory and the practices and findings of modern science.

Here is a second point of conflict with modern science. Russell gives no account of how the properties of ephemerons give rise to the generally recognized properties of electrons and protons, such as mass and electric charge. Russell acknowledges these properties, although perhaps a bit grudgingly:

> It is clear that, in some sense, there are electrons and protons, and we cannot well doubt the substantial accuracy of their estimated masses and electric charge. That is to say, these constants evidently represent something of importance in the physical world, though it would be rash to say that they represent exactly what is at present supposed. (Russell 2007, 394)

However, he makes no attempt to relate the mass or electric charge of electrons and protons to properties of the "metaphysically more primitive" ephemerons.

There are two things that a theorist could say here. One is that the ephemerons are the ultimate bearers of the mass and electric charge that physicists ascribe to electrons and protons. A good example of this approach is the handling of electric charge in today's quark theory of protons and neutrons. According to the quark theory, a proton contains two up-quarks, each with a charge of +2/3, and one down-quark with a charge of -1/3, giving the proton a net charge of +1. Similarly, a neutron contains one up-quark with a charge of +2/3 and two down-quarks, each with a charge of -1/3, giving the neutron a net charge of 0. Alternatively, a theorist could say that the "metaphysically more primitive" ephemerons are bearers of certain "metaphysically more primitive" properties that give rise in some specified way to the mass and electric charge of electrons and protons.

Russell says neither of these things; the story he tells about ephemerons makes no mention of either mass or electric charge. Moreover, his descriptions of ephemerons give no hint of how something made of ephemerons could have properties such as mass and electric charge. He never mentions any difference between the ephemerons that constitute electrons and the ephemerons that constitute protons. He never even indicates that there is a difference. There is one passage in which he distinguishes "luminous events," which constitute light, from events "of the kinds associated with matter" (Russell 2007, 373–374), but he never says what "the kinds associated with matter" are. The only properties of ephemerons that he ever mentions are the ones that I have already acknowledged—size, shape, duration, and causal ties to other ephemerons—plus "intrinsic character." I discuss "intrinsic character" in the next chapter. Here I need only note that Russell says nothing about the "intrinsic character" of ephemerons that suggests a way of relating it to the mass and electric charge of electrons and protons.

One commentator on Russell describes ephemerons as "very short segments of space-time" (Slater 1999, 17). This description is not right; Russell clearly says that ephemerons *occupy* small regions of space-time, not that they *are* small regions of space-time. However, I can sympathize with Slater's remark, because Russell's meager descriptions of ephemerons do not differ much from descriptions of empty regions of space-time. His ephemerons are mere specks of occupancy, if that makes any sense. Such a threadbare world seems incapable of even having mass, electric charge, and other generally recognized properties of electrons and protons. In the name of an "analysis of matter" Russell seems to ignore the generally recognized properties of matter and invent something else for the word "matter" to denote—causally connected specks of occupancy with intrinsic character.

Imagine that you and a friend are seated at a table on which there is a freshly baked loaf of bread. Your friend maintains that the loaf is divided into quarter-inch slices. You see no evidence of cuts in the loaf and you find his arguments unconvincing. Still, you might dismiss the issue as unimportant, thinking that a loaf is a loaf, whether it is sliced or not. But suppose in addition that your friend always refers to the slices as "chapters" and never mentions that these "chapters" have any of the generally recognized properties of bread, such as being tasty and nutritious. Now his claim that the loaf is sliced takes on a troubling aspect. It starts to seem like a device for denying the basic nature of bread. Russell's presentation of his ephemeron theory is very much like this, it seems to me. Just as chapters in the everyday sense are not tasty and nutritious, so events in the everyday sense do not have properties like mass and electric charge. By slicing electrons and protons in the temporal dimension and calling the resulting short-lived entities "events," Russell manages to remove from consideration many of the generally recognized properties of electrons and protons instead of doing the important work of accounting for them.

<>

In the following passage, Russell applies his ephemeron theory to human brains:

> What is in the brain by the time the physiologist examines it if it is dead, I do not profess to know; but while its owner was alive, part, at least, of the contents of his brain consisted of his percepts, thoughts, and feelings. Since his brain also consisted of electrons, we are compelled to conclude that an electron is a grouping of events, and that, if the electron is in a human brain, some of the events composing it are likely to be some of the "mental states" of the man to whom the brain belongs. Or, at any rate, they are likely to be parts of such "mental states"—for it must not be assumed that part of a mental state must be a mental state. I do not wish to discuss what is meant by a "mental state"; the main point for us is that the term must include percepts. Thus a percept is an event or a group of events, each of which belongs to one or more of the groups constituting the electrons in the brain. (Russell 2007, 320)

223

The idea here, as I understand it, is that the ephemerons that constitute a living human brain somehow manage to form groups of two different kinds. Groups of one kind consist of ephemerons that exist one after another in temporal succession, and some of these constitute electrons, protons, and neutrons. Groups of another kind consist of ephemerons that exist simultaneously, and some of these constitute percepts, thoughts, and feelings that exist during that short time interval. This is a bit like saying that the elements of a two-dimensional matrix are grouped into rows and also into columns. Just as a matrix element belongs to a row and to a column, so an ephemeron can belong to an electron and also to a visually experienced patch of purple, for example. This is not to say that Russell thinks a living brain is structured exactly like a matrix. The analogy is merely that a living brain, like a matrix, has ultimate constituents that form subgroups of two different kinds. Between a matrix and the elements that are its ultimate constituents, there are on the one hand rows and on the other hand columns. Between a living brain and the ephemerons that are its ultimate constituents, there are on the one hand electrons, protons, and neutrons and on the other hand percepts, thoughts, and feelings.

Something peculiar about the passage I just quoted is its failure to mention sleep. A significant fraction of a human life is spent in dreamless sleep. Therefore, Russell's statement that "while its owner was alive, part, at least, of the contents of his brain consisted of his percepts, thoughts, and feelings" is at best a very rough approximation. This is not an isolated omission. To my knowledge, Russell *never* discusses sleep. He always writes as if a person is continuously awake from birth to death. Here are some other strangely "sleepless" passages:

> ... our percepts and "mental states" are among the events which constitute the matter of our brains. (Russell 2007, 322)

> ... my head consists of my percepts and other similar events ... (Russell 2007, 382)

> ... each region of the brain is a class of events, and among the events constituting a region thoughts are included. It is to be observed that if we say thoughts are in the brain, we are using an ellipsis. The correct statement is that thoughts are among the events which, as a class, constitute a region in the brain. A given thought, that is to say, is a member of a class, and the class

is a region in the brain. In this sense, where events in brains are concerned, we have no reason to suppose that they are not thoughts, but, on the contrary, have strong reason to suppose that at least some of them are thoughts. I am using "thoughts" as a generic term for mental events. (Russell 1948, 230)

To make room for sleep, one must, at a minimum, make an adjustment to Russell's view. The brain consists in part of percepts, thoughts, and feelings "whenever its owner has percepts, thoughts, and feelings" one would have to say, and "I do not profess to know what is in the brain during that part of each day when its owner is in a state of dreamless sleep." This raises a fundamental question. How does it happen that the ephemerons that constitute a living human brain sometimes do and sometimes do not form groups that are percepts, thoughts, or feelings? An advocate of the ephemeron theory should have a better response than "I do not profess to know." Yet the theory seems to have no resources with which to fashion a better response.

What is even more peculiar about Russell's discussion of the brain is that he never mentions its usual scientific description in terms of neuroanatomy, electrical activity, neurotransmitters, and so on. Just as he describes ephemerons without relating them to the usual scientific description of electrons and protons, he presents an ephemeron-based description of the brain without relating it to the usual scientific description of the brain. The reader is thus left on his own to ponder a host of questions about the relation between Russell's ephemeron-based description of the brain and the usual scientific description of the brain. Are the two descriptions compatible? If there are conflicts, which description is correct? Does Russell's ephemeron-based description of the brain add anything credible to the usual scientific description?

I will answer the last question first. Russell's description of the brain does add something credible to the usual scientific description, namely that the brain includes its owner's inwardly observable experience. The usual scientific description of the brain is based entirely on outward observation and therefore says nothing about inwardly observable experience. As far as outward-observation-based scientific descriptions of the brain are concerned, inwardly observable experience might belong to the brain, or it might be something additional to the brain, or it might not exist at all. Russell declares firmly for the first alternative: a living human brain includes experience and is to some extent inwardly observable. For reasons given in previous chapters, I agree with Russell on this important point.

This is the extent of my agreement with Russell's account, however. The trouble is that his ephemeron-based description of the brain is difficult if not impossible to reconcile with the usual scientific descriptions, which enjoys strong evidential support. Following are four points of conflict between Russell's description of the brain and the usual scientific description.

First, Russell's failure to account for the electric charge of electrons and protons in terms of the properties of ephemerons has a macroscopic echo in his discussion of brains. According to modern science, the brain is made up of molecules and cells that cohere and function in large part because of the electrical interactions between electrons and protons. According to Russell, the brain consists of "groups" or "classes" of ephemerons—terms that suggest mere coexistence and temporal succession without any interaction. Does Russell mean to imply that the usual scientific description is wrong in this fundamental respect? Does he propose that we should abandon the basic findings of neuroscience in favor of his theory? That seems like a hopeless position to take. But if that is not his position, then his talk of "groups" and "classes" is misleadingly incomplete. At a minimum, it needs to be replaced by a description of a network or system of electrically interacting ephemerons.

Second, Russell's failure to account for the mass of electrons and protons in terms of the properties of ephemerons also has a macroscopic echo in his discussion of brains. According to modern science, a human brain weighs about three pounds. Does Russell mean to imply that the brain consists entirely of weightless ephemerons and hence has no weight? That seems like a hopeless position to take. On the other hand, if he grants that the brain weighs about three pounds, then he must say that its constituent ephemerons have weight and therefore that percepts, thoughts, and feelings have weight. That is not obviously false, but it is a surprising claim that calls for explicit acknowledgment and defense. Note that the quadrants view does not have this implication because it says that the brain of an awake or dreaming human being includes percepts, thoughts, and feelings as properties, not as parts. Percepts, thoughts, and feelings do not have weight, according to the quadrants view, for the same reason that the brain's temperature does not have weight. Russell's key idea that percepts, thoughts, and feelings are parts of the brain, made of the same ephemerons that the brain as a whole is made of, forces him to choose between the brain being weightless and percepts, thoughts, and feelings having weight.

Third, Russell's description of the brain has a numbers problem. If a human body contains more than 10^{27} electrons, as I estimated in chapter 2, then a human brain, which contributes more than 1% of the body's weight, contains more than 10^{25} electrons. Given Russell's view that electrons and protons are really sequences of short-lived ephemerons, the brain at any moment must consist of more than 10^{25} ephemerons. Neuroscientists estimate that the brain contains 100 billion (10^{11}) neurons. It follows that an average neuron at a given moment must consist of more than 10^{14}, or 100 trillion, ephemerons. In light of these huge numbers, Russell's suggestion that a percept, thought, or feeling is a group of some small number of ephemerons—possibly even a single ephemeron—seems incredible. Which few electrons and protons out of all these billions and trillions are the ones whose constituent ephemerons also constitute a person's percepts, thoughts, and feelings? Denying that the brain contains this huge number of electrons and protons seems like a hopeless position. To avoid this consequence, one could modify Russell's description to say that a typical percept, thought, or feeling is a huge group of perhaps 10^{20} ephemerons. This would be a major change. A key feature of Russell's view is that inward observation reveals the nature of the brain's ultimate constituents. One can see this theme in the following passages that span several decades:

> Since the immediate data of sense are not indestructible but in a state of perceptual flux, it is argued that these data themselves cannot be among the ultimate constituents of matter. I believe this to be a sheer mistake. The persistent particles of mathematical physics I regard as logical constructions, symbolic fictions enabling us to express compendiously very complicated assemblages of facts; and, on the other hand, I believe that the actual data in sensation, immediate objects of sight or touch or hearing, are extramental, purely physical, and among the ultimate constituents of matter. (Russell 1959, 128)

> So long as it is supposed that the physical world is composed of stable and more or less permanent constituents, the fact that what we see is changed by changes in our body appears to afford reason for regarding what we see as not an ultimate constituent of matter. But if it is recognized that the ultimate constituents of matter are as circumscribed in duration as in spatial extent, the whole of this difficulty vanishes. (Russell 1959, 134)

> I contend that the ultimate constituents of matter are not atoms or electrons, but sensations, and other things similar to sensations as regards extent and duration. (Russell 1921, 121)

> Or perhaps, since the subject matter of physics is no longer matter in the old sense, it may be that what we call our thoughts are ingredients of the complexes with which physics has replaced the old conception of matter. (Russell 1931, 126–127)

> But there is one part of the physical world which we know otherwise than through physics, namely that part in which our thoughts and feelings are situated. These thoughts and feelings, therefore, are members of the atoms (or minimum material constituents) of our brains. (Russell 1963, 706)

It is not easy to reconcile these passages with the idea that inward observation encounters groups of perhaps 10^{20} ephemerons. Such a huge number suggests that a human being's inward observation reveals nothing about the ultimate constituents of matter, even if the ultimate constituents are ephemerons of the sort that Russell describes. According to the quadrants view, inward observation is observation of certain properties of a complex electronuclear system. These properties are of course related to the ultimate constituents of the system, but extremely distantly.

Fourth, Russell's description of the brain is difficult to reconcile with the apparent fact that living human beings are experientially very different from other objects. If percepts, thoughts, and feelings are small groups or classes of ephemerons, then it would seem that percepts, thoughts, and feelings much like ours should exist wherever there are assemblages of electrons and protons—in a potato or a mud puddle, for example. Although Russell sometimes seems open to a kind of panpsychism, the following passages are not compatible with the idea that percepts, thoughts, and feelings of the human kind exist wherever electrons and protons congregate:

> On empirical grounds we believe that there cannot be visual objects except where there are eyes and nerves and a brain ... (Russell 1940, 354–355)

> I think that experience is a very restricted and cosmically trivial aspect of a very tiny portion of the universe. (Russell 1995a, 50)

It is difficult to suppose that sticks and stones feel either pleasure or unpleasure... (Russell 1995a, 183)

How can percepts, thoughts, and feelings like ours be so rare if they are small groups or classes of ephemerons, which are so abundant? There is a dissonance here that Russell seems not to recognize. The natural way to avoid this dissonance is to say that features of human experience are properties of certain extremely complex systems and that their existence depends on that complexity.

In several passages, Russell makes the odd-sounding claim that brain scientists cannot "see" other people's brains:

Whoever accepts the causal theory of perception is compelled to conclude that percepts are in our heads, for they come at the end of a causal chain of physical events leading, spatially, from the object to the brain of the percipient....

It follows from this that what the physiologist sees when he examines a brain is in the physiologist, not in the brain he is examining. (Russell 2007, 320)

I should say that what the physiologist sees when he looks at a brain is part of his own brain, not part of the brain he is examining. (Russell 2007, 383)

It is natural to suppose that what the physiologist sees is in the brain he is observing. But if we are speaking of physical space, what the physiologist sees is in his own brain. (Russell 1995b, 110)

What you see when you look at a brain through a microscope is part of your private world. It is the effect in you of a long causal process starting from the brain that you say you are looking at.... What I maintain is that we can witness or observe what goes on in our own heads, and that we cannot witness or observe anything else at all. (Russell 1995a, 18–19)

To understand these passages correctly, it is necessary to be aware that Russell uses the word "see" much more narrowly than most people do. In Russell's terminology, all one ever "sees" is one's own visual experience—which he calls visual "percepts." More generally, all one ever "perceives" is one's own sensory experience. From sensory experience, visual or otherwise, he says that people "infer" things about the objects from which the spatial transmissions come. Given Russell's usage, for example, no one can see dogs or automobiles for the same reason that a brain scientist cannot see other people's brains. People "see" visual representations of dogs and automobiles, from which they "infer" that dogs and automobiles having various characteristics are out there in front of them. Russell thus uses the word "infer" very broadly, in a way that includes not only the various forms of reasoning for which the word is usually reserved, but also much that most people call seeing or perceiving.

Given that Russell uses the words "see" and "infer" in this way, I think he is correct that everything a brain scientist sees when he studies another person's brain is in his own brain. I question the wisdom of using the word "see" in this unconventional way, but I agree with Russell's point. The trouble is that this point fails to address certain crucial questions. Framed in Russell's terminology, these questions are:

- What can brain scientists *infer* about brains from the perceptual experiences that they have while they are studying brains? For example, can they infer that brains contain billions of neurons having mass, electrical activity, complex branching shapes, and synaptic connections?
- How are the properties of brains that brain scientists can infer from their perceptual experiences related to inwardly observable experiential properties?

The many passages in which Russell insists that a brain scientist can "see" only his own brain seem suspiciously like attempts to dodge these questions. One might even take these passages to suggest that a brain has no outwardly observable properties—the polar opposite of the more common suggestion that brains have only outwardly observable properties. That would provide quick answers to both questions. Unfortunately, it is wildly implausible. It would be absurd to throw out all the findings of modern neuroscience in order to make room for Russell's ephemeron-based description of the brain. If the two cannot be reconciled and integrated with each other, it is Russell's theory that has to go.

<>

In sum, I see no reason at all to give credence to Russell's ephemeron theory of electrons, protons, and brains. His main argument that electrons and protons are movie-like sequences of static ephemerons, which is based on relativity theory, is fallacious. The many decades of laboratory work since Russell advanced this theory have produced no evidence that electrons and protons have such a staccato structure. Russell's suggestion that laboratory work is not relevant to the theory is mysterious; it is also at odds with his reliance on the argument from relativity theory. His ephemeron theory has no resources for explaining widely recognized properties of electrons and protons such as mass and electric charge. Likewise, the idea that the brain is a four-dimensional swarm of ephemerons has no resources for explaining the brain's mass, anatomy, or electrical and chemical activity. The claim that inwardly observable thoughts, feelings, and percepts are small groups of simultaneously existing ephemerons is hard to reconcile with the huge number of electrons and protons in the brain. It is also hard to reconcile with the absence of thoughts, feelings, and percepts in potatoes, mud puddles, and the brains of sleeping people, where there should be similar small groups of simultaneously existing ephemerons according to the theory. Russell's avoidance of all discussion of the brain's outwardly observable properties as described in contemporary neuroscience is suspicious and strange. His "analysis of matter" and his description of the brain simply ignore much of what scientists think they have learned about electrons, protons, and brains. Although Russell always speaks of science with respect, he seems willing to ignore scientific findings that do not fit well with his theory.

There is one point discussed in this chapter concerning which I think Russell is importantly correct: a person who is inwardly observing his percepts, thoughts, and feelings is observing his brain. He is not observing an independently existing "thinking thing" or puffs of "mental steam" that his brain emits. But this point is easy to separate from the ephemeron theory. According to the quadrants view, the brain is not a four-dimensional swarm of ephemerons small groups of which constitute percepts, thoughts, and feelings. Rather, it is a complex electronuclear system certain properties of which are percepts, thoughts, and feelings. Russell's ephemeron theory of electrons, protons, and brains is fundamentally wrong.

17

The Quadrants and Bertrand Russell: Inference from Perception

I noted in the previous chapter that Russell uses the words "see" and "perceive" with unconventionally narrow meanings and the word "infer" with an unconventionally broad meaning. Most people say that they see and perceive nearby objects, but in Russell's usage a person sees and perceives only features of his own experience, from which he makes inferences about his surroundings. In Russell's usage, for example, people do not see dogs or automobiles. They "see" only visual representations of dogs and automobiles, from which they "infer" that dogs and automobiles having various characteristics are out there in front of them.

Within his broad category of inference from perception, Russell draws the following distinction. The basic awareness of one's surroundings that most people call seeing or perceiving Russell calls "physiological inference" (Russell 2007, 50, 190–193). For inference that involves explicit reasoning he uses several terms, including "deliberate scientific inference" (Russell 2007, 178), "explicit inference," "valid inference by means of a true principle" (Russell 2007, 190), and "conscious inference" (Russell 2007, 219). He uses these terms casually and not necessarily synonymously, but they all contrast with "physiological inference" in that they involve explicit reasoning. I'm puzzled by the way Russell uses the phrase "physiological inference." A distinction between inference that is automatic or spontaneous and inference that is thoughtful or methodical strikes me as being neutral with respect to physiology, and it seems to me that it should strike Russell that way too. In any case, Russell is aware

232

that his broad category of inference from perception includes both spontaneous processes and methodical processes.

Inference from perception understood in Russell's broad way coincides with learning from outward observation, as I am using that term. Accordingly, one can directly compare what Russell says about inference from perception with what I say about outward observation. That comparison is the main business of this chapter.

<>

One important difference is that Russell seeks to characterize in a comprehensive way everything that people can reasonably infer from perception, whereas I simply say that objects have outwardly observable properties. I don't reject the possibility that all outwardly observable properties might have something in common in addition to being outwardly observable, but I don't see any additional commonality and I see problems with Russell's claims that he does see one. One of these problems is that Russell seems to offer two or three different characterizations of the scope of inference from perception without acknowledging the differences. There is a jitter in his discussion of the subject that is reminiscent of his shifting descriptions of the type of structure that is supposed to relate ephemerons ("events") to electrons and protons. A second and more serious problem is that there are many apparent counterexamples to all of Russell's characterizations, including several counterexamples mentioned by Russell himself. A detailed examination of these problems follows.

Russell's account includes descriptions of 1) the processes that produce sensory experience, 2) the content of sensory experience, and 3) the processes by which people make inferences from sensory experience. I consider each of these in turn.

Russell's main claim about the processes that produce sensory experience is sometimes called "the causal theory of perception," a term that he applies to his view. The claim is that sensory experience is caused by incoming spatial transmissions that strike the sense organs. I agree with this statement, provided that it is accompanied by the recognition that sensory experience also has causes in the observer's body. When I open my eyes, for example, the colorful panorama of visual experience that springs into existence results from the *convergence* of incoming electromagnetic radiation with certain parts and properties of my nervous system. It is not the case that incoming spatial transmissions simply turn into sensory experience upon arrival at a certain location.

Russell's main claim about the content of sensory experience is that it has both "structure" and "intrinsic quality" or "intrinsic character." He uses the phrases "intrinsic quality" and "intrinsic character" interchangeably. Sometimes I use both of these terms together, and sometimes I use only "intrinsic character."

Russell explains in detail and in several places that he means something specific by "structure." The structure of something, in Russell's sense, consists of relations that obtain between and among the thing's parts (2007, 249–256; 1948, 250–256; 1995a, 72–76). Something with many parts that bear many relations to one another has a complex structure, in Russell's sense, while something with no parts has no structure. This is an important stipulation because the meaning of the word "structure" in everyday usage is extremely broad and nebulous. There are things that one can call aspects of the world's structure without offending anyone's sense of proper English that do not qualify as structure in Russell's sense. In fact, there is little if anything that cannot be considered an aspect of the world's structure, in a broad and nebulous everyday sense. To avoid this ambiguity of the word "structure," I am going to introduce the term "R-structure" for structure in Russell's sense. You can think of the "R" as standing for "Russell" and/or "relation."

Russell's handling of "intrinsic quality" and "intrinsic character" is importantly different from his handling of "structure." He does not define these terms at all, but only gives examples of what he takes them to include. Because he is mainly concerned with vision, the examples he uses most often are color sensations—experiential green, experiential red, and so on. But there are many comparable examples for the other senses. Experiential green, experiential red, the experience of hearing a piano key struck, the experience of hearing a violin string vibrate, the experience of smelling perfume, the experience of smelling smoke, the experience of tasting a peach, the experience of tasting chocolate, the experience of holding your hand under cool running water, the experience of holding your hand under a warm blow dryer—each of these and countless other features of sensory experience has its own distinctive "intrinsic character," in Russell's terminology.

Russell's main claim about inference from perception is that one can make reasonable inferences from the "structure" of sensory experience but not from its "intrinsic quality" or "intrinsic character." He uses the view known as naïve realism as a foil for his own view. A person in the grip of naïve realism unthinkingly attributes the intrinsic character of his own sensory experience to the objects that play a role in causing that sensory experience. For example, he supposes that the trees on a distant hillside are experiential green, that during a thunderstorm the atmosphere is filled with experiential rumbling, and that a ripe peach hanging on a tree is shot through with experiential peach flavor. These spontaneous inferences of naïve realism are wrong, according to Russell (and according to most people who have thought about the question), because all these features of experience are produced in the perceiving individual. It comes naturally to people to attribute features of their sensory experience to the objects that play a role in causing the experience, but this is an unreasonable inferential procedure because it is based on a misconception.

The only reasonable inferences from perception, Russell says, are based on the R-structure that consists of the spatial and temporal relations between contrasting bits of experiential intrinsic character. He makes this point, using experiential color as an example, when he says:

> … we may suppose that what is happening in a place is what a person would perceive in that place, provided we use, in inference, only those properties of the percept which it shares with the stimulus. E. g. we must not use the blueness of blue, but we may use its difference from red or yellow. (Russell 2007, 227)

The R-structure of sensory experience is a reasonable basis for inference, according to Russell, because it resembles the R-structure of the distant objects that play a role in causing sensory experience. The relation between the R-structure of sensory experience and the R-structure of its causes is like the relation between a map and the region it maps (Russell 2007, 249). There is support for this idea in the fact that similar R-structure often shows up in different sensory modalities, such as visual and tactile encounters with the same object or a spatial sequence of musical notes on a page and a temporal sequence of heard musical notes. There is additional support in studies of the spatial transmissions that show their capacity to possess and preserve R-structure in their travels through the vast expanse.

As a simple illustration of these ideas, consider what you can learn about a house by looking at it from the street. The bricks are not experiential red; the mortar is not experiential white; the shutters are not experiential black. To believe such things is to fall victim to naïve realism. On the other hand, there is a door with windows on each side of it and there are shutters on each side of each window. This is part of the R-structure of the house, which you

235

can reasonably infer from the R-structure of your visual experience because the causal process that produces your visual experience maps R-structure from the house into your visual experience.

I accept Russell's critique of naïve realism. I also agree with him that we can make reasonable inferences from the R-structure of our sensory experience. However, I do not agree that R-structure is the *only* basis for reasonable inferences from perception. Recall that Russell never defines the "intrinsic character" of sensory experience, but only gives examples of it. Consequently, his two master categories—"R-structure" and "intrinsic character"—are not logically exhaustive. There is logical room for aspects of sensory experience that are not covered by either term. And there are several aspects of sensory experience that these two categories may not cover. They are not cases of "R-structure" and they seem rather different from Russell's favorite examples of "intrinsic character." Whether or not they are additional examples of "intrinsic character" is indeterminate and inessential. What matters is that it is very difficult to classify them as "R-structure" and very plausible that they can be used to make reasonable inferences about the world. Accordingly, they seem to be counterexamples to Russell's central claim that the only reasonable basis for inference from perception is R-structure. After discussing several of these apparent counterexamples, I will assess their significance.

One apparent counterexample is the spatial character of visual experience, by which I mean the magnificent billowing breadth and depth of it. When I open my eyes, it's as if my brain blows a gigantic bubble within which all my visual representations of objects take up their positions. This feature of my visual experience, along with my successful use of it as a guide for all that I do with my arms and legs, is the basis of my belief that I am in a vast expanse. The billowing breadth and depth of my visual experience is a property of my waking body that models the vast expanse that my body is in. A second apparent counterexample is the temporal character of experience in general, which includes the temporal character of sensory experience. This is the basis of my belief that the world at large has a temporal character. As I noted above, Russell handles experiential space and time as spatial and temporal relations that belong to the R-structure of sensory experience. He is certainly correct that there are spatial and temporal relations in sensory experience. They include being above and below, to the left of and to the right of, behind and in front of, before and after, near and far away. But the point he seems to miss is that these relations are not simply instances of relatedness. Spatial relations have a distinctive character that is different from the character of temporal relations, and this is because spatial relations are grounded in the nonrelational spatial character of experience while temporal relations are grounded in the nonrelational temporal character of experience. Whether or not one counts the spatial and temporal character of sensory experience as part of its "intrinsic character," it is a basis of reasonable inference from perception that is not R-structure.

A third apparent counterexample is shape. Russell clearly believes that sensory experience provides a basis for making reasonable inferences about the shapes of objects. He creates an appearance of harmony between this belief and his general claim that all reasonable inference concerns R-structure by using examples of shapes made of planes, straight lines and angles that can be considered interrelated parts:

> ... the structural properties of an object seen are approximately identical with those of the same object touched. If you have models of the regular solids, the one which you can see to be a dodecahedron will be correctly named by an educated blind man after feeling it. (Russell 1948, 226–227)

> If you see something hexagonal, then, since hexagonality is a structural property, the physical object which has caused your visual sensation must be hexagonal, although its hexagonality will be in a space which is not identical with visual space. (Russell 1948, 468)

This sounds good, but what about the smoothly curving shape of a polka dot, a pear, or an eggplant? In general, shape is not a matter of relations between parts, either in sensory experience or in the world at large. There is an area of overlap between shape and R-structure, occupied by dodecahedrons and hexagons, but there are also shapes that do not consist of R-structure, such as polka dots, pears, and eggplants. Russell seems to miss this point.

A fourth apparent counterexample is motion of all kinds, including motion along a path, rotation, and vibration. The sensory basis for inference to motion is multi-faceted. It includes smooth directional progressions that take place within the billowing breadth and depth of visual experience. It includes the sense of what is happening to your body when you walk, change posture, or turn around. And it includes changes in visual and tactile experience that are coordinated with your bodily actions. According to Russell's ephemeron theory, there is no real motion, but only swarms of static short-lived ephemerons in space-time. Presumably, therefore, he considers all inferences to motion mistaken, although I don't believe he says this explicitly. However, if the ephemeron theory is wrong, as I argued in the previous chapter, then it is reasonable to infer from perception that objects are moving in various ways, and so the apparent counterexample of motion stands.

A fifth apparent counterexample is recognition. There is recognition of recurrent aspects of sensory experience, such as a melody or the taste of a common item in your diet. There is recognition of individual electronuclear objects, such as a person or the Statue of Liberty. There is recognition of object types, such as horses or maple leaves. Recognition of individual objects and object types that are not part of your sensory experience seems to involve an inference, in Russell's broad sense, from a purely sensory recognition, and this is not a mere matter of projecting R-structure. For example, recognizing that the object in front of you is a horse depends on having a purely sensory horse-recognition of the sort that you can also have when you look at a painting or photograph of a horse.

Russell's neglect of sensory recognition may be related to his emphasis on the incoming causes of sensory experience, which I mentioned earlier. Sensory recognition requires a causal contribution from within that brings to bear memories of previous experience. Russell often likens the human visual system to a camera. There are similarities between visually capable human beings and cameras, but there are also differences. One important difference is that cameras do not recognize things.

Recognition is also an apparent counterexample to Russell's claim that there is no role for intrinsic character as such in reasonable inference from perception: "we must not use the blueness of blue, but we may use its difference from red or yellow." It is reasonable to use the blueness of blue to recognize bluebirds. It is reasonable to use the intrinsic character of a speaker's voice to recognize the speaker. Russell was presumably thinking of naïve realism when he stated this principle, but the statement is overly broad because it disallows the reasonable recognition of individual objects and object types based on the intrinsic character of the sensory experience that they help produce. Attributing the intrinsic character of sensory experience to distant objects is a mistake, but using the intrinsic character of sensory experience to recognize distant objects is not a mistake.

Recognition of individual objects and object types is an indispensable prerequisite for still other important sorts of inference from perception. You need recognition in order to build up a body of belief about a particular object or object type. Over time, you augment and revise what you believe about something through a series of encounters with it, which you recognize to be repeat encounters with the same thing. Without recognition, you would not be able to combine the inferences that you make on one occasion with the inferences that you make about the same object or object type on another occasion. This can be considered a sixth apparent counterexample. Recognition also enables us to count things. One always counts members of a recognized kind—galaxies or neurons or chickens, for example. This is true of estimated counts as well as exact counts. Much of statistical inference depends on counting things. Counting and all that depends on counting can be considered a seventh apparent counterexample to Russell's claim that the sole basis of reasonable inference from perception is the R-structure of sensory experience.

<>

I turn now to a second way in which Russell characterizes the scope of inference from perception. In addition to claiming that reasonable inference from perception can tell us only about "structure," Russell claims that it can tell us only about "mathematical properties":

> ... our knowledge of physics is mathematical ... because no non-mathematical properties of the physical world can be inferred from perception. (Russell 2007, 253)

> The only legitimate attitude about the physical world seems to be one of complete agnosticism as regards all but its mathematical properties. (Russell 2007, 270)

> By examining our percepts it is possible ... to infer certain formal mathematical properties of external matter, though the inference is not demonstrative or certain. (Russell 2007, 382)

Russell writes as if the claim that one can infer only "structure" and the claim that one can infer only "mathematical properties" are equivalent, yet it is hard to see how these claims can be equivalent unless he is using words in a highly unusual way. There are many relational facts that are not normally considered mathematical. Where is the mathematics in the claim that the lungs are inside the rib cage and alongside the heart? Conversely, there seem to be many mathematical facts that are not relational. Counting and the statistics of counting are a rich source of examples. The claim that there are over seven billion living human beings is mathematical if numerical facts are mathematical, yet it involves no relation.

It seems that when Russell speaks of "mathematical properties" he is thinking mainly of the mathematical *equations* of science, but even here the link to R-structure is dubious. It is true that a written mathematical equation itself has R-structure: the variables, constants, and operators that it consists of are its parts and the relations between these parts are indicated by the way all these symbols are arranged on the page. But the property of the world that the equation expresses is another matter. Take as a simple example the law that relates the volume, pressure, and temperature of the gas in a closed container. The R-structure of the gas concerns the spatial and temporal relations of the jostling gas molecules, but this is not what the mathematical equation expresses. The equation relates three measurable quantities—pressure (P), volume (V), and temperature (T):

$$PV = nRT$$

Moreover, inferring the validity of this equation involves building measuring instruments, using the instruments to take measurements, performing the computations indicated by the expressions on each side of the equation, and checking that the result of computing PV and the result of computing nRT are close enough, given the imperfections of the measuring instruments. In short, it involves much more than the inferential projection of R-structure from sensory experience to its distant causes.

The following passage, which Russell uses to introduce a discussion of the work of Darwin and Pavlov, seems to contradict his own claim that "no nonmathematical properties of the physical world can be inferred from perception":

> Perhaps in its ultimate perfection all science will be mathematical, but in the meantime there are vast fields to which mathematics is scarcely applicable, and among these are to be found some of the most important achievements of modern science. (Russell 1931, 40–41)

This is a much more sensible statement. It seems to me that science—and cognition in general—are chock-full of inferences from perception that do not concern "mathematical properties." Russell is simply wrong that it is a limitation in our capacity for inference from perception that "accounts for the fact that our knowledge of physics is mathematical." This should be obvious from the fact that physics is *more* mathematical than many other fields of study that are just as dependent on inference from perception. If there is an explanation for the prominence of mathematics in physics, it must be different from the one that Russell gives. One possible explanation is that the subject matter of physics is especially well-suited to mathematical treatment. Another is that mathematically-trained physicists tend to look at the world through "mathematics-colored glasses" and therefore pay more attention to what they can express mathematically.

In addition, Russell's underlying claim that "our knowledge of physics is mathematical" is wrong in the sense that he intends it. True enough, physicists use a lot of mathematics. But according to Russell, our knowledge of physics is *exclusively* mathematical, and that is not the case. Every mathematical equation that physicists use contains variables that represent measurable quantities. All the quantities that go into these equations are determined by subjecting various nonmathematical properties to measurement procedures. The example of the gas law illustrates this point nicely. The law is expressed by a mathematical equation, but behind that equation there are the nonmathematical properties of pressure, temperature, and spatial expanse for which measurement procedures have to be devised and carried out. The mathematical aspect of physics is just that—an aspect. A mathematical equation cannot have descriptive validity unless its variables are tied to nonmathematical properties for which we have measurement procedures. Moreover, measurable nonmathematical properties such as spatial expanse, temperature, pressure, mass, gravitational force, electric force, and so on are not examples of R-structure. They are apparent counterexamples to the general claim that reasonable inference from perception leads only to mathematical properties and they are also apparent counterexamples to the general claim that reasonable inference from perception leads only to R-structure.

Russell claims that force is "a mere term in a mathematical equation" and "a mathematical fiction" (Russell 2007, 161). I don't know what this means exactly, but it seems far from true. I believe that the perceptions one has while trying to lift a heavy piece of furniture or separate two magnets point to real nonmathematical properties of the world. Statements like these make we wonder whether Russell's thinking is based in part on the illusion of blackboard physics. If you walk into a physics lecture toward the end, you'll see a blackboard filled with equations and diagrams—lots of mathematics and R-structure. But that is not the whole of what physicists infer from perception. All those chalk marks have ties to encounters with the world that go beyond mathematics and R-structure—recognizing the objects that one is studying, discovering properties that can be measured, building instruments to measure those properties, using the instruments to make measurements, and so on. The writing on the blackboard is only part of an edifice that includes many other reasonable inferences from perception. Russell began his career as a pure mathematician; perhaps this led him to think of physics as more like pure mathematics than it actually is.

In the following passage, Russell goes a step further and claims that inference from perception is limited not only to mathematical properties, but to mathematical laws:

> We know the laws of the physical world, in so far as these are mathematical, pretty well, but we know nothing else about it. (Russell 2007, 264)

This is plainly false because the basis for belief in a law is belief in many particular facts that fit the pattern that the law specifies. Moreover, people have been able to infer from perception many particular facts for which no lawful pattern has been found. Considering the fruits of inference from perception as a whole, laws are the exception, not the rule. Elsewhere, Russell acknowledges as much, contradicting his general claim:

> On the whole, the tendency of the foregoing discussion has been to suggest that it is easy to exaggerate the evidence for simple laws in the physical world. Where we know most—i.e. in regard to the structure of the atom—there is, so far as we know a complete absence of law in certain very important respects. (Russell 2007, 236)

This is another example of Russell contradicting himself without realizing it. In the course of a long book, he presents several different general principles about inference from perception and apparent counterexamples to all of them.

In my view, all these apparent counterexamples to Russell's general claims involve inferences from perception that are just as reasonable as the inference of R-structure or "mathematical properties" from the R-structure of sensory experience. Russell's preoccupation with R-structure seems arbitrarily neglectful of other important aspects of sensory experience and inference from sensory experience. I conclude that reasonable inference from perception cannot be characterized in any of the monolithic ways that Russell proposes, and that the actual harvest of inference from perception is not as meager as Russell maintains. Outward observation is a more complex, more varied, and more fruitful enterprise than Russell makes it out to be.

It is probably not a coincidence that Arthur Eddington, writing at the same time as Russell, presents a similarly restrictive view about the scope of outward observation and the physical sciences. Where Russell speaks of "structure" and "mathematical properties," Eddington speaks of "pointer readings":

> The whole subject matter of exact science consists of pointer readings and similar indications. (Eddington 1958, 252)

> The recognition that our knowledge of the objects treated in physics consists solely of readings of pointers and other indicators transforms our view of the status of physical knowledge in a fundamental way. (Eddington 1958, 258)

> The physical atom is, like everything else in physics, a schedule of pointer readings. (Eddington 1958, 259)

The pointer readings are said to be "attached to some unknown background" (Eddington 1958, 259), which is made of "mind-stuff" according to the panpsychist speculations discussed in chapter 12. If this was a fair description of physics, the theories that physicists produce would be mere works of pointer-reading art that had no claim to credibly represent the reality that gives rise to the "pointer readings." The fact is that the pointer readings do not come forth from "some unknown background." The world that gives rise to the pointer readings is partially unknown, of course, but it is far from being completely unknown. People have designed and manufactured the measuring instruments that produce the pointer readings, and the reason people have made these measuring instruments is to better understand various properties that they observe outwardly but non-numerically. The numerical pointer readings are only a part of the relevant fund of outward observation, and they are not even the most basic part. Reasonable inference from perception is not restricted to R-structure or to "mathematical properties" or to "pointer readings." All these claims fail to acknowledge the full scope and variety of outward observation.

<>

I will now describe some similarities and differences between Russell's view and the quadrants view. These comparisons bring to light additional problems of Russell's view.

According to Russell, sensory experience is a combination of R-structure and "intrinsic quality" or "intrinsic character." He applies this same dichotomy to the world at large: *everything* is a combination of R-structure (or perhaps "mathematical properties") and intrinsic character. In the case of your own sensory experience, the R-structure and the intrinsic character are both available for you to inwardly observe. In the case of objects in general—acorns and water molecules, for example—a person who studies them can make reasonable inferences about their R-structure and/or "mathematical properties" from the R-structure of his sensory experience, but he cannot make any inferences about their intrinsic character. Intrinsic character in the world at large is something that Russell says human beings have no way to learn about.

There is a formal parallel between this characteristic of Russell's view and the quadrants view. According to the quadrants view, objects such as acorns and water molecules have outwardly observable properties, which human beings can learn about, and remote properties, which human beings cannot learn about. According to Russell, objects such as acorns and water molecules have R-structure and/or mathematical properties, which human beings can learn about, and intrinsic character, which human beings cannot learn about. I think Russell is correct to draw a line between what we can and cannot learn about such objects through outward observation. But I also think that he draws the wrong line in the wrong way. According to Russell, the scope of inference from perception is limited by the nature of the perceptual and inferential processes in the human observer who receives the spatial transmissions. According to the quadrants view, the scope of inference from perception is limited by the mismatch between the evidence-carrying competence of the spatial transmissions and the nature of the electronuclear fabric at the point of origin of the spatial transmissions. According to the quadrants view, there is no property of incoming light or any other sensory stimulus from which one could infer the remote properties of electronuclear objects, so the nature of the perceptual and inferential processes that these stimuli feed into is beside the point. I don't rule out the possibility that human perceptual and inferential processes impose some additional limitation on what human beings can infer from the spatial transmissions, but I see no reason to think that this is the case. Remember that we can use any feasible technology to pre-process these transmissions and manipulate the stimuli that strike our sense organs.

This contrast between the quadrants view and Russell's view resembles a contrast between the quadrants view and McGinn's view that I discussed in chapter 15. McGinn thinks that human beings do not have the ability to understand how their experience is generated, but that a differently constituted being might have this ability. According to the quadrants view, no electronuclear being can understand how its experience is generated because all are equally cut off from the remote properties of electronuclear objects, for the same exogenous reason. Unlike McGinn, Russell does not speculate about the cognitive powers of possible nonhuman beings. He does not ask whether his division of sensory experience into R-structure and intrinsic character applies beyond human beings, and if so how far beyond. Like McGinn, however, Russell focuses on internal cognitive processes, leaving exogenous factors out of consideration. Both Russell and McGinn maintain a tight focus on cognitive processing within observers, which is a major difference between their theorizing and the quadrants view.

A major problem with Russell's view is that it seems to imply the possibility of a type of inference from perception that is in fact not possible. Russell never asks whether there is any limit to a person's ability to make inferences about the world's R-structure from the R-structure of his sensory experience. Although he does not say that all of the world's R-structure lies open to human inferential powers, there is nothing in his theory to suggest that some of the world's R-structure is cognitively closed to us. That fact together with the general tenor of his discussion gives me the impression that he thinks of the human ability to infer the world's R-structure as in principle unlimited. Obviously, some inferences to R-structure will be especially difficult, such as inferences to R-structure at extremely small scales of space and time or inferences about the R-structure in the center of the sun, but Russell lays down no absolute limits.

From this it seems to follow that from the R-structure of one's own sensory experience one can in principle infer the R-structure of the sensory experience of all other living things that have sensory experience. For example, one should be able to infer the R-structure of the sensory experience of an echo-locating bat. One could not infer the intrinsic character of the bat's sensory experience, of course, because one cannot infer the intrinsic character of anything, but one should be able to arrange a perceptual encounter with a living bat such that the R-structure of the bat's sensory experience is mapped into the R-structure of your own sensory experience. Based on Russell's account, one might even expect that inferring the R-structure of another animal's sensory experience should be especially easy, since it is the R-structure of your own sensory experience that is the basis of the whole enterprise of inference from perception.

None of this is the case, however. The R-structure of the sensory experience of other animals is not accessible in the way that the outwardly observable R-structure of houses, trees, and automobiles is. Moreover, Russell seems to accept this. He concludes only cautiously that other animals have sensory experience at all, and he makes no pretense of being able to infer the details of its R-structure. As I noted in the last chapter, he thinks that experience is a part of brains, and he avoids the subject of making perception-based inferences about brains altogether. In chapter 14 I summarized three erroneous explanations of the fact that one cannot outwardly observe a person's experiential properties—1) because that person can inwardly observe them (Flanagan), 2) because they have a "subjective mode of existence" (Searle), and 3) because they are not in the space in which outward observation takes place (McGinn). Russell's theory of inference from perception has the opposite problem of implying that this is not a fact. According to his account, inferring the R-structure of another person's experiential properties should be an exercise on a par with inferring the R-structure of a boulder or a building.

Earlier in this chapter I explained why Russell's principle of inference from the R-structure of sensory experience rules out many examples of seemingly reasonable inference from perception. I now conclude that it also *rules in* a type of inference from perception that is seemingly not possible, namely inference to the R-structure of another animal's sensory experience. Russell's principle is fundamentally misconceived, like a medical test that gives many correct results but also many false negatives and many false positives. This conclusion is consistent with the quadrants view, because the definition of the quadrants has nothing to do with R-structure. According to the quadrants view, one can reasonably infer a lot of R-structure from perception, and that is the extent of the relation between the two. Russell twists this fact into a congruence thesis by focusing selectively on examples of inferable R-structure. Examples of inferable properties other than R-structure and examples of R-structure that is not inferable get swept under the rug.

Yet another problem with Russell's view is that R-structure and Russell's "intrinsic character" do not seem to fully cover the nature of the world, just as they do not seem to fully cover sensory experience. If intrinsic character in the world at large is something that we have no way to infer from perception, then the union of R-structure and intrinsic character leaves out many inferable properties that do not fall under R-structure. Examples include spatial expanse, smoothly bounded shapes, motion, mass, temperature, and the forces. In someone else's terminology, all these properties might count as aspects of the world's intrinsic character, but they are excluded by Russell's condition that intrinsic character is not inferable from perception. This incomplete coverage of the world at large by Russell's categories dovetails with two problems that I discussed in the previous chapter—his failure to account for the generally recognized properties of electrons and protons in terms of the properties of ephemerons, and his avoidance of the question of what brain scientists can infer about brains. On all these fronts, Russell's description of the world seems to leave a lot out. Instead of plenary reality, he gives us a quasi-mathematical object enhanced with intrinsic character.

A fourth and final problem is that Russell repeatedly suggests that there is a *resemblance* between the inwardly observable intrinsic character of human sensory experience and the intrinsic character of the world at large, but he never says what he thinks the resemblance consists in. If he took an explicit stand for panpsychism, he could maintain that the intrinsic character of all ephemerons is experiential. He does not do this, however, and he seems to think that the resemblance he alludes to can obtain between experiential intrinsic character and nonexperiential intrinsic character. The trouble is that I cannot think of any possible respect of resemblance between experiential intrinsic character and nonexperiential intrinsic character. Russell's suggestion that there is some resemblance between the intrinsic character of the world at large and the intrinsic character of human sensory experience therefore seems empty.

Following are several passages in which Russell makes this resemblance claim. After each passage, I ask a question for which there seems to be no answer.

> In regard to what happens to ourself, we know not only abstract logical structure, but also qualities—by which I mean what characterises sounds as opposed to colours, or red as opposed to green. This is *the sort of thing* that we cannot know where the physical world is concerned. (Russell 1959, 19; my emphasis)

What *sort* of thing is this? I cannot think of a category that includes these features of sensory experience and also something nonexperiential.

> The gulf between percepts and physics is not a gulf as regards intrinsic quality, for we know nothing of the intrinsic quality of the physical world, and therefore do not know whether it is, or is not, *very different from that of percepts*. (Russell 2007, 264; my emphasis)

> As to intrinsic character, we do not know enough about it in the physical world to have a right to say that it is *very different from that of percepts* ... (Russell 2007, 400; my emphasis)

If intrinsic character in the world at large is not "very different from that of percepts," there must be some similarity. What could the similarity be, if being experiential is excluded?

> We now realise that we know nothing of the intrinsic quality of physical phenomena except when they happen to be sensations, and that therefore there is no reason to be surprised that some are sensations, or to suppose that the others are *totally unlike sensations*. (Russell 1995b, 117; my emphasis)

If the physical phenomena that are not sensations are not "totally unlike sensations," there must be some similarity. What could the similarity be, if being experiential is excluded?

> ...we have no reason to assert that the events in us are *so very different from the events outside us*—as to this, we must remain ignorant, since the outside events are only known as to their abstract mathematical characteristics, which do not show whether these events are like "thoughts" or unlike them. (Russell 1995b, 170–171; my emphasis)

If "the events in us" are not "so very different from the events outside us" or if "the events outside us" are "like 'thoughts'," there must be some similarity. What is the similarity, if being experiential is excluded?

There is no theoretical reason why a light-wave should not consist of groups of occurrences, each containing a member *more or less analogous* to a minute part of a visual percept. (Russell 2007, 263; my emphasis)

If the one is "more or less analogous" to the other, what is the analogy?

In sum, Russell finds many ways to say that there is a resemblance between the intrinsic character of sensory experience and the intrinsic character of the world at large, but no way to say what the resemblance is. To fill in this blank, one needs to specify a possible respect of resemblance. I cannot think of any way to do this, unless one adopts panpsychism. It therefore seems to me that all these resemblance claims are empty.

What gives Russell the idea that the intrinsic character of his sensory experience can resemble the intrinsic character of something that has no experience? I think part of the answer is that Russell holds the ephemeron theory that I criticized in the previous chapter. Believing that each feature of his sensory experience is made of a small number of ephemerons encourages the belief that his inward observation of his sensory experience gives him a kind of insight into the nature of the fundamental building blocks of all things. He thus arrives at the nebulous notion that the ephemerons that constitute his kitchen table are not very different from, not totally unlike, and more or less analogous to the ephemerons that constitute his sensory experience.

There are two important contrasts between this aspect of Russell's view and the quadrants view. First, the quadrants view does not include any claim of positive resemblance between experiential properties and remote properties. According to the quadrants view, remote properties and experiential properties are alike in not being outwardly observable. Positive resemblances between all or some experiential properties and all or some remote properties are not ruled out, but they are not part of the view. Second, according to the quadrants view all features of experience are properties of extremely complex electronuclear systems of the distinctive sort that we call animals, so there is no reason to think that your inward observation of your experience gives you any insight at all into your kitchen table. You do resemble your kitchen table in some very general ways:

- Both you and your kitchen table are made entirely of electronuclear fabric.
- Both you and your kitchen table have properties that are outwardly observable and other properties that are not outwardly observable.

But your electronuclear make-up has all sorts of complexities that your kitchen table does not have. Because of this difference, you have no reason to think that your inward observation of your experience gives you any insight into the nature of your kitchen table or any of its components.

18

The Quadrants and Daniel Stoljar

Daniel Stoljar devotes his book *Ignorance and Imagination* to a problem that he calls "the logical problem of experience." This problem concerns a set of three statements that he labels T1, T2 and T3, as follows:

> T1. There are experiential truths.
>
> T2. If there are experiential truths, every experiential truth is entailed by some nonexperiential truth.
>
> T3. If there are experiential truths, not every experiential truth is entailed by some nonexperiential truth.
>
> (Stoljar 2006, 26, 67, 218)

"T" stands for "thesis." Stoljar claims that "each of the theses has powerful considerations, or what seem initially to be powerful considerations, in its favor" (2006, 26). However, it is impossible for all three to be true because their conjunction is a logical contradiction. I think there is room for debate about how powerful or even powerful-seeming the considerations are that Stoljar cites in support of these three statements. My own view is that the case he makes for T1 is much stronger than anything he says in support of either T2 or T3. However, the problem can be posed regardless of the strength of these considerations, simply because the conjunction of T1, T2, and T3 is a logical contradiction. Since these three statements cannot all be true, the challenge is to say which one is false and to explain why.

As you can see, all three statements are extremely terse. Before trying to decide which one is false, it is a good idea to examine in some detail what they say.

Let's start with the word "truths." Stoljar says nothing about this word, apparently regarding it as self-explanatory. And perhaps it is. But there is a characteristic of this word that I must call attention to because it plays an important role throughout this chapter. A truth is true, and in order to be true it must be a thing of a kind that can be true. This means, I believe, that it must be a thing of a cognitive kind. Some examples are beliefs, hypotheses, conjectures, statements, propositions, and sentences. Plants and pebbles, colors and shapes, cannot be truths because they do not have the cognitive character that is a prerequisite for being true. Suppose we introduce the term "truth candidate" to cover everything that is of a kind that can be true. Then we can say that a truth is a truth candidate that is true.

Next is the all-important distinction between "experiential truths" and "nonexperiential truths." Stoljar explains this distinction as follows, apparently using the pronoun "we" to refer to himself:

> ... when we speak of experiential truths, what we mainly have in mind are truths with a certain distinctive subject matter— namely, events of sensory, perceptual, and imaginative experience, events whose defining property is that there is something it is like to undergo them. Likewise, when we speak of nonexperiential truths, what we have in mind are truths with a certain, different, distinctive subject matter— namely, events that are not experiences. (Stoljar 2006, 26)

This explanation is too simple. It overlooks the fact that there are many truths whose "distinctive subject matter" is neither purely experiential nor purely nonexperiential, but rather a relationship between something experiential and something nonexperiential. There are truths about how sense organs help to produce sensory experience. There are truths about how various diseases and injuries help to produce experiential symptoms. There are truths about how artistic performances such as concerts, plays, movies, magic shows and comedy routines help to produce the experiential effects that the performers strive for. There are truths about how various features of experience are related to outward observations of the brain such as electrical recordings or brain images. There are truths about how various features of experience prompt, guide, or shape behavior. Relationships between the experiential and the nonexperiential are the subject of much everyday curiosity as well as much scientific research. Accordingly, it is important to specify whether these mixed truths belong to the category of "experiential truths" or the category of "nonexperiential truths."

The answer has to be that all truths that have a mixture of experiential and nonexperiential subject matter count as experiential truths. This is the only classification that makes the problem interesting. It is corroborated toward the end of the book when Stoljar cites "Fred's arthritis causes him pain" and "Looking at Granny Smith apples causes green sensations" as examples of experiential truths (2006, 202). Any truth that mentions experience or some particular feature of experience counts as an experiential truth, no matter how many references to nonexperiential objects or properties it contains. Only truths that do not mention experience at all count as nonexperiential truths. This point is of such importance that I think it is a good idea to reword the three statements to make it explicit:

T1 (reworded). There are truths that mention experience.

T2 (reworded). If there are truths that mention experience, then every truth that mentions experience is entailed by some truth that does not mention experience.

T3 (reworded). If there are truths that mention experience, then not every truth that mentions experience is entailed by some truth that does not mention experience.

These statements are slightly longer than those discussed by Stoljar, but the additional length is justified by the gain in clarity.

Here is a second revision of the problem description. Stoljar says that T1 is "an empirical claim of quite astounding obviousness," citing as an example of such a truth "I am having experiences right now" (2006, 26). I agree; T1 is not a serious contender for the statement that is false. Given this, there is no need to keep T1 in the problem description. We can drop T1 along with the "if" clauses in T2 and T3 that link them to T1. The problem then boils down to a choice between these two statements:

T2 (reworded and then simplified). Every truth that mentions experience is entailed by some truth that does not mention experience.

> T3 (reworded and then simplified). Not every truth that
> mentions experience is entailed by some truth that does not
> mention experience.

Equivalently, the problem is to give and explain the correct answer to the
following true/false question. Every truth that mentions experience is entailed
by some truth that does not mention experience—true or false?

The phrase "is entailed by" also calls for discussion. Throughout his book,
Stoljar speaks of entailment as a relation between truths. Strictly speaking,
this is not correct. Entailment is a relation between *truth candidates* that does
not depend on whether they are true or not. Stoljar gives as an example of
entailment that "Snow is white" entails "Something is white" (Stoljar 2006,
34). This is a good example, but considered in isolation it can create the
false impression that entailment only obtains between truths. One needs an
assortment of examples involving both true truth candidates and false truth
candidates in order to paint a correct picture. "Snow is red" entails "Something
is red"; here a falsehood entails a truth. "All integers less than 20 are prime
numbers" entails "All integers less than 10 are prime numbers"; here a falsehood
entails a falsehood. The case of a falsehood entailing a falsehood is extremely
important, not a mere curiosity. It is the backbone of the *reductio ad absurdum*
method of mathematical proof, in which one proves that a statement is false by
showing that it entails a logical contradiction, which is necessarily false. The
key point is that entailment is a cognitive relation between truth candidates
that depends on their cognitive content, but not on whether their cognitive
content is true or false. What matters is whether the cognitive content of one
truth candidate or a certain set of truth candidates "intelligibly determines"
(Stoljar 2006, 101) the cognitive content of another truth candidate. Whether
this relation of intelligible determination obtains between two truth candidates
is one question, whether the truth candidates are true or false is another.

<>

Stoljar approaches the logical problem of experience in a way that
introduces unnecessary difficulties. He defines an abstract object that he calls
"the experiential conditional" in the following way:

> ... suppose ... that all the nonexperiential truths that in
> fact obtain are conjoined into one large truth (call it N),

and all the experiential truths are conjoined into another large truth (call it E).... Now consider the truth-functional conditional formed by N and E ("If N, then E"), and call this "the experiential conditional." (Stoljar 2006, 35)

He then says that T2 is true if and only if the experiential conditional is an entailment (N entails E). Relying on this equivalence, he focuses his attention on the all-encompassing experiential conditional instead of on separate entailments of many individual experiential truths, as T2 and T3 do. This is a dubious strategy for two reasons.

First, either N or E or both of them might be infinite conjunctions, and this possibility casts a shadow over much of what Stoljar says about them. He imagines a hypothetical individual who knows N, knows E, and uses this knowledge to answer the question whether a relation of intelligible determination obtains between N and E (Stoljar 2006, 155–160). None of this makes sense if either N or E is infinite. He seems to assume that the set of experiential truths and the set of nonexperiential truths are both finite sets, but he says nothing to justify this assumption.

Stoljar is not alone in making such a finiteness assumption, but having company is not the same thing as having a justification. In chapter 9, I noted that Frank Jackson provides no justification for his belief that research in the physical sciences can lead to "a complete account of non-sentient reality." One of the arguments for T3 that Stoljar criticizes is an adaptation of Jackson's widely discussed "knowledge argument," which I mentioned in chapter 14. Jackson describes Mary, the supremely knowledgeable scientist who has been deprived from birth of all color experiences except black, white, and shades of gray, in the following way:

> Mary is confined to a black-and-white room, is educated through black-and-white books and through lectures relayed on black-and-white television. In this way she learns everything there is to know about the physical nature of the world. She knows all the physical facts about us and our environment, in a wide sense of "physical" which includes everything in *completed* physics, chemistry, and neurophysiology, and all there is to know about the causal and relational facts consequent upon all this.... (Jackson 2004b, 51; his emphasis; also quoted in Stoljar 2006, 38)

It is possible—indeed likely, in my opinion—that the world is such that physics, chemistry, and neurophysiology admit of limitless elaboration. If this is the case, there is no such thing as "*completed* physics, chemistry, and neurophysiology," and Jackson's use of this phrase makes his argument dependent on an overly simple conception of the world. Although Stoljar is highly critical of Jackson's knowledge argument on other grounds, he seems to accept this aspect of it. David Papineau seems to make the same unjustified finiteness assumption when he speaks of "a full physical description of the world" (Papineau 2002, 153). There can be no *full* physical description of the world if the physical sciences admit of limitless elaboration.

The second problem with the experiential conditional is that there is a very plausible idea, which Stoljar seems to accept, that a given individual's ability to understand experiential truths is limited by the range of that individual's experience. This opens up the possibility that statement E includes truths concerning such a variety of features of experience belonging to such a variety of organisms and expressed in such a variety of languages that no one individual can understand them all. For example, E might include truths about the experience of intelligent beings from another planet that no human can understand and truths about human experience that no extraterrestrial can understand. There is room for an incomprehensible variety even within the ambit of human experience. E might include truths about the experience of primitive tribes that no modern human can understand and truths about the experience of modern humans that no primitive human can understand. E might include truths about the experience of males that no female can understand and truths about the experience of females that no male can understand. E might include truths about the experience of autistic people that only autistic people can understand and truths about the experience of nonautistic people that no autistic person can understand. And so on. This possibility casts a shadow over Stoljar's discussion of an individual who knows and reflects on E, even if E is finite.

It seems to me that both of these difficulties are avoidable because there is no need for Stoljar to introduce the experiential conditional. He could have kept his focus on T2 and T3, asking whether each and every individual experiential truth is entailed by some finite conjunction of nonexperiential truths. As far as I can tell, every passage in his book that mentions the experiential conditional can be recast so as to make a parallel point about a large and possibly infinite set of entailments, where each individual entailment involves only truth candidates of finite length. Accordingly, I am going to set the experiential conditional aside and discuss Stoljar's reasoning in this parallel form.

<>

What, then, is the solution to Stoljar's logical problem of experience? Stoljar maintains that the correct statement is T2: Every truth that mentions experience is entailed by some set of truths that do not mention experience. I think the correct statement is T3: Not every truth that mentions experience is entailed by some set of truths that do not mention experience. Moreover, I think T3 is an enormous understatement. In my view, it is probably the case that *no* truth that mentions experience is entailed by a set of truths that do not mention experience. If there are exceptions to this generalization, I would expect them to be rare and weird. On the question of whether entailments of this sort exist, Stoljar and I are poles apart.

In making his case for T2, Stoljar pursues two related tasks without clearly distinguishing them. Included in his description of the logical problem of experience is a certain family of arguments that others have made for T3. One of these arguments is an adaptation of Frank Jackson's knowledge argument. There are also several so-called conceivability arguments. In defending his choice of T2 over T3, Stoljar devotes many pages to criticizing these arguments. He even devotes many pages to criticizing certain published criticisms of these arguments, on the ground that they do not correctly diagnose where the arguments go wrong. Criticism of existing arguments for T3 is a major theme of his book. So far, so good; as an advocate of T2, Stoljar needs to explain what is wrong with any known argument for T3. But here is the problem. Stoljar is so preoccupied with criticizing these arguments for T3 that he does not make clear the distinction between showing that these arguments are flawed and showing that their conclusion, T3, is false. His presentation creates the impression that discrediting this family of arguments for T3 is tantamount to discrediting T3 itself. This confusion is manifest in statements such as the following:

> Moreover, if we are making one of these mistakes, the arguments for T3 collapse and the logical problem is solved. (Stoljar 2006, 79)

Not so fast! It is a commonplace of elementary logic that one can construct fallacious arguments that have a true conclusion. A person can reject all of the arguments for T3 that Stoljar criticizes, or be unsure what to make of those arguments, and yet choose T3 over T2 for other reasons. I am such a person. I do not endorse any of the arguments for T3 that Stoljar criticizes, but I think there is good reason to choose T3 over T2. My reason for choosing T3 is based on the quadrants view.

Ironically, my reason for choosing T3 is rather similar to Stoljar's reason for choosing T2. Of course, the reasons are different enough to lead to opposite solutions to the problem, but they are similar nevertheless. The centerpiece of Stoljar's reason for choosing T2 is what he calls "the ignorance hypothesis." I will refer to this hypothesis as "Stoljar's ignorance hypothesis" or "his ignorance hypothesis" in order to keep it distinct from the different ignorance hypothesis that is integral to the quadrants view. To solve the logical problem of experience *correctly*, one must understand the similarities and the differences between these two ignorance hypotheses.

Stoljar states his ignorance hypothesis in several slightly different ways, sometimes using the word "ignorant" and sometimes using the word "unaware":

> ... we are ignorant of a type of experience-relevant nonexperiential truth. (Stoljar 2006, 6, 67)

> ... we are unaware of a type of experience-relevant nonexperiential truth. (Stoljar 2006, 68–69)

> ... we are unaware of a type of nonexperiential truth relevant to the nature of experience—an experience-relevant nonexperiential truth. (Stoljar 2006, 87)

> ... we are ignorant of or unaware of a type of truth relevant to the nature of experience. (Stoljar 2006, 113)

The truths posited by Stoljar's ignorance hypothesis are experience-relevant in the sense that they make good the entailments that exist according to T2:

> In the sense at issue here, one truth T is relevant to a truth T* just in case T is an essential part of a set of truths that together entail T*. To illustrate, consider the argument "All men are mortal; Socrates is a man; therefore, Socrates is mortal." The premise "All men are mortal" is relevant to the

conclusion but does not entail it; rather, it is an essential part of a set of truths that does entail the conclusion—namely, the set that contains "All men are mortal" and "Socrates is a man." When I say that we are ignorant of an experience-relevant truth, I mean that we are ignorant of a truth that stands to experiential truths in much the same way. (Stoljar 2006, 70)

Stoljar's use of the two words "ignorant" and "unaware" reflects a subtlety in his ignorance hypothesis that is easy to miss on a first reading. It might seem that his ignorance hypothesis says simply that there is a certain type of truth no example of which is known to be true. But actually, it says this and more. It also says—using my terminology—that there is a certain type of *truth candidate* no example of which has come to people's attention. It is not the case that we can formulate lots of entailments of the sort that would vindicate T2, but we just don't know whether their premises are true. The situation, rather, is that we are not able to formulate any entailments of the sort that would vindicate T2. To emphasize this point, one could say that Stoljar's ignorance hypothesis posits two layers of ignorance. The first and deeper layer is unawareness of the type of truth candidate that would make it possible to formulate entailments of the sort that vindicate T2. The second layer, which is implicit in the first, is ignorance of the truth of any truth candidate of that unknown type.

To help explain his ignorance hypothesis, Stoljar uses the example of a congenitally blind person's unawareness of color experience (2006, 69). The two layers of ignorance are apparent in this example. A congenitally blind person cannot appreciate the truth of a statement such as "When electromagnetic radiation with a wavelength of 470 nanometers strikes the retina of a living and awake human being, it triggers the generation of blue color experience." But equally, a congenitally blind person cannot appreciate the falsehood of a statement such as "When X-rays strike the retina of a living and awake human being, they trigger the generation of blue color experience." A congenitally blind person does not know any truths about blue color experience because he does not understand any truth candidates that mention blue color experience. Stoljar's claim is that all the entailments that vindicate T2 exist, but we can neither formulate them nor know that all their premises are true because some of their premises are of a type that we are unaware of in something like the way that a congenitally blind person is unaware of blue color experience.

Stoljar's ignorance hypothesis has an abstract, bare-bones character—a point that he emphasizes. All it says is that there are truths of an unknown type that make good the entailments that vindicate T2. It says nothing about what distinguishes the truth candidates of this unknown type from truth candidates of known types. It says nothing about the subject matter of the truth candidates of this unknown type. It says nothing about why this unknown type of truth candidate remains unknown at this point in the history of human inquiry. And it says nothing about the prospects for human beings someday coming to learn truths of this unknown type. Noting McGinn's conjecture that human beings are "cognitively closed" to certain things that cognitively more competent beings can understand, Stoljar says that human beings might or might not be "cognitively closed" to the unknown type of truth that his ignorance hypothesis posits. "It may be that these truths could not be expressed in a language that we could speak or understand," he writes (Stoljar 2006, 35). But on the other hand, he says that "they might be truths that are knowable by us" (Stoljar 2006, 94). All these questions are left completely open by Stoljar's ignorance hypothesis. Of course, these questions must have answers if Stoljar's ignorance hypothesis is true, but Stoljar need not provide the answers in order to make a case for his bare-bones ignorance hypothesis.

I said earlier that Stoljar does not make clear the distinction between discrediting a certain family of arguments for T3 and discrediting T3 itself. I think he is led to conflate these two tasks because his ignorance hypothesis plays a central role in both of them. He seems not to realize that it plays a different role in each task. To discredit the arguments for T3, it is enough to show that his ignorance hypothesis describes a possibility that those arguments do not rule out. I think he does this successfully; his ignorance hypothesis describes a possibility that those arguments do not even consider. On the other hand, to discredit T3 he needs to make a convincing case that his ignorance hypothesis is true. This is a higher bar, and he does not clear it.

<>

According to the quadrants view, we are ignorant of the remote properties of electronuclear objects. These remote properties, or at least some of them, are "experience-relevant" in the sense that they play a crucial role in the generation of experiential properties; experiential properties are jointly caused by outwardly observable properties and remote properties. We cannot understand the causation of experiential properties because we are unaware of a certain type of property that plays a crucial role in it. Stoljar holds that every "experiential truth" is entailed by a set of "nonexperiential truths" that includes some truths of an unknown type. We cannot formulate or understand these entailments because we are unaware of a certain type of truth candidate that plays a crucial role in them. Stoljar's ignorance hypothesis leaves open whether the nonexperiential truths of an unknown type entail experiential truths all by themselves or in conjunction with nonexperiential truths of a known type. It is thus compatible with the idea that experiential truths are jointly entailed by nonexperiential truths of a known type and nonexperiential truths of an unknown type. *Joint entailment* of experiential truths in accordance with Stoljar's ignorance hypothesis corresponds to *joint causation* of experiential properties in the quadrants view.

There are three main differences between these two ignorance hypotheses.

First, Stoljar's ignorance hypothesis says nothing specific about the unknown type of truth, whereas the quadrants view includes certain specific claims about the remote properties of electronuclear objects. Stoljar's ignorance hypothesis is silent about the unknown type of truth's distinguishing characteristics, subject matter, basis of current elusiveness, and potential for being discovered in the future. According to the quadrants view, on the other hand, the reason that some properties of electronuclear objects have the extreme cognitive remoteness of remote properties is the mismatch between the nature of electronuclear fabric and the spatial transmissions on which all outward observation depends. And because this reason for ignorance applies to all feasible electronuclear beings in the vast expanse, ignorance of the remote properties of electronuclear objects is the permanent condition of all feasible electronuclear beings.

Second, the one commitment that Stoljar's ignorance hypothesis does make is that there are experience-relevant *truths* of an unknown type. According to the quadrants view, on the other hand, the reason for our ignorance of the remote properties of electronuclear objects has the further consequence that there no truths regarding particular remote properties. Because these properties have no evidentiary effects on the spatial transmissions on which outward observation depends, they have no representation in the thought of any feasible electronuclear being. If thought is exclusively an electronuclear phenomenon, it follows that the remote properties of electronuclear objects can have no representation in thought, period. They are an aspect of reality remote from all possible cognition. The ignorance posited by Stoljar's ignorance hypothesis is ignorance of truths. The ignorance posited by the quadrants view is ignorance of an aspect of reality concerning which there are no truths.

The third difference is a consequence of the first two. The two ignorance hypotheses lead to opposite solutions to Stoljar's logical problem of experience. According to Stoljar's ignorance hypothesis, there are nonexperiential truths of an unknown type that make good the entailments that vindicate T2. According to the quadrants view, every experiential property that can be mentioned by an "experiential truth" is causally dependent on a combination of outwardly observable nonexperiential properties concerning which there are truths and remote nonexperiential properties concerning which there are no truths. There is joint causation of experiential properties by outwardly observable properties and remote properties, but there is no corresponding joint entailment of experiential truths because remote properties and their causal role in the generation of experience have no representation in thought. If there are any "experiential truths" that are entailed by a set of "nonexperiential truths" that do not concern remote properties, these would be weird and aberrant cases. In general and perhaps without exception, the entailments that would vindicate T2 do not exist. Stoljar's ignorance hypothesis, which posits truths of an unknown type, supports T2. The ignorance hypothesis that is part of the quadrants view, which posits unknown properties concerning which there are no truths, supports T3.

<>

Which of these ignorance hypotheses is more plausible? Stoljar makes several arguments for the plausibility of his ignorance hypothesis, which I examine below. There is nothing in any of them that weakens the case for the quadrants view that I presented in previous chapters.

Stoljar says that his ignorance hypothesis gains plausibility from its complete openness regarding the nature of the truths of an unknown type. This is a reasonable-sounding application of the principle that the less you say, the smaller is your risk of saying something wrong. However, Stoljar's inability to say anything at all about the truths of an unknown type is actually a significant weak point in his case. He says that he has no reason to believe that human beings can understand the truths of an unknown type; maybe they can and maybe they can't. What he does not point out is that he gives no reason to believe that electronuclear beings of any sort can understand the truths of an unknown type. Like McGinn, he gives no way to rule out or even to make implausible the possibility that all feasible electronuclear beings are cognitively limited in a way that undermines his hypothesis. Further, Stoljar gives no reason to believe that there is an omniscient being that can understand the truths of an unknown type. If there is no feasible electronuclear being and no omniscient being that can understand the truths of an unknown type, there are no truth candidates of this unknown type and hence no truths. Stoljar's crucial claim that we are ignorant of *truths* that make good a class of *logical entailments* is undercut by his inability to establish the feasibility of a being who can entertain these truths.

A second plausibility argument appeals to the ideas of Bertrand Russell that I discussed in the previous two chapters. Stoljar says that his ignorance hypothesis is consonant with Russell's more specific ignorance hypothesis, which has some plausibility in Stoljar's opinion. One issue here is how plausible Russell's view is; see my criticisms of it in chapters 16 and 17. But what is more germane to the logical problem of experience is the fact that Stoljar appeals to a "Russellian view" that differs from Russell's actual view in a crucial respect. What Russell says, in brief, is that the world is made of tiny, short-lived, static "events" and that there is something about our human powers of perception and inference that limits what physical scientists are able to learn about these "events." Physical scientists are able to discover "mathematical" or "structural" properties of things made of "events" but they are not able to discover the "intrinsic character" or "intrinsic quality" of the "events." With the exception of the intrinsic character of our own experience, therefore, we are ignorant of the world's intrinsic character. In the "Russellian view" employed by Stoljar, ignorance of intrinsic character is gratuitously described as ignorance of *truths about intrinsic character*. I see nothing in Russell's account to suggest that he thinks of human ignorance of intrinsic character as ignorance of truths about intrinsic character. As I noted in chapter 17, Russell does not ask whether there could be nonhuman electronuclear beings whose powers of perception and inference put them in touch with the world's intrinsic character. And Russell definitely does not believe in an omniscient being. But if there is no feasible being, electronuclear or otherwise, that can understand truths about the intrinsic character of nonexperiential "events," then there are no truth candidates of this type and hence no truths. One can reasonably read Russell's descriptions of human ignorance of the intrinsic character of "events" as descriptions of ignorance of an aspect of reality concerning which there are no truths. If you read Russell's descriptions in this way, Russell's work does not support Stoljar's ignorance hypothesis.

A third plausibility consideration that Stoljar cites concerns certain historical arguments, one made by Descartes and one by C. D. Broad. Stoljar sees a structural similarity between these arguments and the arguments for T3 that he criticizes. Originally, the historical arguments seemed persuasive, but the subsequent discovery of new types of relevant truths—"computational truths" in the case of Descartes's argument and "truths about quantum mechanics" in the case of Broad's argument—made it apparent that the arguments were flawed and that their conclusions were false. The structurally similar arguments for T3 fit the same pattern, Stoljar claims. The distinction between discrediting certain arguments for T3 and discrediting T3 is important here. As I noted earlier, the arguments for T3 that Stoljar criticizes do not rule out the possibility that Stoljar's ignorance hypothesis is true. This alone shows that they are inconclusive. Historical examples of structurally similar arguments that were made in ignorance of relevant truths and that have false conclusions serve to drive this point home. But citing these historical examples as a reason to reject T3 is a gratuitous generalization, akin to arguing "These two dogs are brown, therefore all dogs are brown." It's perfectly possible that some faulty arguments of this sort, such as the historical arguments of Descartes and Broad, have a false conclusion, while others, such as the arguments for T3 that Stoljar criticizes, happen to have a true conclusion.

Stoljar says that his ignorance hypothesis says "nothing at all about the limits of thought" (Stoljar 2006, 11). This is not so. It is true that his ignorance hypothesis makes no claim that thought *is limited* in any way. But his ignorance hypothesis does say, implicitly, that thought *is not limited* in the way that the quadrants view says it is. If thought is limited in the way that the quadrants view says it is, then Stoljar's solution to the logical problem of experience is wrong. What makes it wrong is precisely the inability of thought—any thought, of any feasible being—to represent certain properties of electronuclear fabric that are causally relevant to experience.

The idea that thought in general is unlimited in its ability to understand the world seems to be an unspoken and unexamined presupposition that pervades Stoljar's book. Expressions such as "the body of truths that describe the world" (Stoljar 2006, 70) and "the complete story about physical objects" (Stoljar 2006, 108) do not make sense in the absence of such a presupposition. Whenever he mentions ignorance, it is always ignorance of truths; he draws no distinction between ignorance of truths and ignorance of an aspect of reality concerning which there are no truths. He writes about the logical problem of experience as if the totality of truths were a proxy for the whole of reality. The possibility that the totality of truths describes only a portion of reality, leaving out another portion that admits of no cognitive representation, receives no consideration.

Stoljar says that his ignorance hypothesis is not committed to any "radical views about the nature and content of the ignorance in question" (Stoljar 2006, 10). He mentions several examples that he considers radical views about ignorance. One sounds like a summary of McGinn's view:

> Some view our ignorance as a consequence of the cognitive structures that are part of our genetic endowment, and that as such we are cognitively closed with respect to the solution to the problem. (Stoljar 2006, 10–11)

Another sounds like a summary of Russell's view:

> And some view our ignorance as a consequence...of the a priori structure of empirical inquiry, as following from the fact (if it is a fact) that empirical inquiry does not acquaint us with the intrinsic nature of matter. (Stoljar 2006, 11)

If Stoljar considers these views radical, then he would presumably say the same about the ignorance claim that is part of the quadrants view: all feasible electronuclear beings are permanently ignorant of certain properties of electronuclear objects because of the mismatch between electronuclear fabric and the spatial transmissions on which all outward observation depends. But is it reasonable for Stoljar to withhold the "radical" stamp from his own ignorance hypothesis? The view that all our ignorance is ignorance of truths seems radical in its own way. Stoljar's ignorance hypothesis is not the modest, minimally committal view that he makes it out to be. It is committed to the strong claim, which conflicts with the quadrants view, that all our experience-relevant ignorance is ignorance of experience-relevant truths.

In any case, whether a view is radical and whether it is right are two different things. For the reasons I have been elaborating over many chapters, I believe that the quadrants view is more right than these other views, wherever it may stand in the "radical" rankings.

<>

The idea that all ignorance is ignorance of truths is wrong in another way that I have not yet discussed. I have been focusing on my claim that the remote properties of electronuclear objects are an aspect of reality concerning which there are no truths, because this claim bears importantly on the solution to the logical problem of experience. But according to the quadrants view the experiential properties of many animals are another aspect of reality concerning which there are no truths. If experiential properties are not outwardly observable, if many animals that have experience have no capacity to inwardly observe their experience or to think about it, and if there is no omniscient being with the capacity to think about the experience of these animals, then the experiential properties of these animals can have no cognitive representation. In a discussion of Nagel's essay "What is it like to be a bat?" Stoljar refers in passing to "the truths, whatever they are, that state what it is like to be a bat" (Stoljar 2006, 155). This is another manifestation of his presupposition that reality in all its aspects is covered by an all-encompassing set of truths. It is easy to doubt that truths that state what it is like to be a bat can be framed in any human language, because human languages are rooted in human experience. It is easy to doubt that bats can state or in any way entertain such truths, because they have little or no capacity for inward observation, thought, or linguistic expression. And it is easy to doubt that there is an omniscient being who can entertain such truths. Therefore, it is easy to doubt that such truths exist at all. If bats have more of the relevant capabilities than I give them credit for, then the point can be illustrated with a cognitively less capable species—toads or termites, for example. Throughout some very substantial portion of the animal kingdom, there are unobservable features of experience concerning which there are no truths.

The existence of features of experience concerning which there are no truths does not affect the *solution* to Stoljar's logical problem of experience, but it does affect the *scope* of the problem. Being about "experiential truths," the logical problem of experience is not about experience in general, but only about that portion of experience concerning which there are truths. In his discussion of the logical problem of experience, Stoljar creates the impression that "experiential truths" are a proxy for the whole of experiential reality. In fact, there is a large portion of experiential reality that plays no role in the problem at all.

<>

The heart of my disagreement with Stoljar concerning his logical problem of experience suggests the following analogy. Let the whole of reality be represented by the whole surface of a planet, and let the portion of reality concerning which there are truths be represented by dry land.

Stoljar presupposes that the relation between reality and the totality of truths corresponds to a Mars-like planet, with a surface consisting entirely of dry land. According to the quadrants view, the relation between reality and the totality of truths corresponds to an earth-like planet, with a surface consisting of multiple land masses surrounded and separated by a global ocean of reality-concerning-which-there-are-no-truths. On the Mars model, all truths are logically interconnected, so the entailments vindicating T2 exist. On the earth model, there can be truths about one land mass that are logically isolated from truths about another land mass. In particular, truths that mention experience can be logically isolated from truths that do not mention experience. The global ocean binds the continents together to constitute a single reality, but it does not connect them logically.

I close this chapter with a four-way comparison of panexperientialism, McGinn's view, Stoljar's view, and the quadrants view. These views have something important in common. According to each of them, the foundation of animal experience includes an element that most writers on the subject fail to acknowledge. The foundation is not to be found entirely in physical-science properties, or at least not in physical-science properties of any known kind. On this point, I believe all these views are correct. However, the views differ concerning the nature of the generally unacknowledged foundational element, as summarized in the following table:

View	Generally Unacknowledged Foundational Element
Panexperientialism	Simple and strange experiential properties of electrons and other fundamental particles
McGinn	Properties of brains that are unintelligible to human beings but intelligible in some nonhuman way
Stoljar	The subject matter of truths of an unknown type, which may or may not be intelligible to human beings
Quadrants	Remote properties of electronuclear objects, which are nonexperiential, intelligible to no feasible electronuclear being, and not the subject of any truths

In this set of four views, the quadrants view stands out for positing the foundational element that has the least resemblance to anything that is widely acknowledged to exist. The simple and strange experiential properties posited by panexperientialism resemble familiar human experience in being experiential. The properties posited by McGinn resemble familiar humanly intelligible properties in being somehow intelligible. Stoljar's truths of an unknown type resemble familiar truths of known types in being truths. But the remote properties posited by the quadrants view have none of these resemblances. They are not experiential, they are not intelligible, and they are not the subject matter of any truths. They have only the minimal resemblance to experiential properties of not being outwardly observable and the minimal resemblance to outwardly observable properties of not being experiential.

This is not a reason to consider the quadrants view more plausible than these other views; I gave those reasons in earlier chapters. It is not a reason to consider the quadrants view less plausible than these other views; familiarity is not an indicator of correctness. I summarize these comparisons simply as one more aid to understanding the quadrants view and its relations to certain views that have been advocated by others.

19

The Quadrants and the Prefix "Proto"

Three of the writers that I discuss in this book use the prefix "proto" to describe views that they defend or at least find plausible. The following table lists these writers and the "proto" terms that they use:

Writer	"Proto" Words Used
Gregg Rosenberg	protoconscious
David Chalmers	protophenomenal
Thomas Nagel	proto-mental, protomental, protopsychic

This chapter discusses the use of this prefix and asks whether there is a useful role for it in describing the quadrants view. My main conclusion is that the prefix "proto" is best avoided in discussions of this subject because different writers use it with significantly different meanings.

Writers on many subjects use the prefix "proto" from time to time. I will start with a general discussion of this prefix, drawing examples from far and wide, and then focus on its use in discussions of the mind-body problem.

<>

Typically, when a writer uses a word of the form "proto-x" the following three conditions obtain:

1. Both "x" and "proto-x" designate stages in a developmental or evolutionary process.

269

2. The proto-x stage precedes the x stage.
3. The proto-x stage resembles and therefore prefigures the x stage.

To illustrate, here are definitions of four "proto" words taken from my dictionary:

+ A protogalaxy is "a cloud of gas believed to be the precursor to a galaxy."
+ A protoplanet is "a hypothetical whirling gaseous mass ... believed to give rise to a planet."
+ A protostar is "a cloud of gas and dust in space believed to develop into a star."
+ A protolanguage is "an assumed or recorded ancestral language."

These examples all satisfy the three conditions. In each case, there is 1) a developmental process that 2) passes through proto-x on its way to x. In addition, 3) proto-x resembles and prefigures x.

Following are four examples of "proto" words that I have come across in recent reading on other subjects. These are *all* the examples that I have encountered while working on this book; I am not cherry-picking examples to fit my analysis.

In *Your Inner Fish*, Neil Shubin describes a 375-million-year-old fossil of a fish that frequented shallow streams and was "capable of doing push-ups" on its fins. Inside each fin was a cluster of small bones that Shubin calls a "proto-wrist" (2008, 36–43). Here we have a developmental process that passes through this fish's cluster of small fin bones on the way to the human wrist. In addition, the cluster of small fin bones resembles and prefigures the human wrist. It thereby qualifies as a proto-wrist.

In *The Ascent of Money*, Niall Ferguson gives a sketch of the history of the board game Monopoly, which includes the following sentence:

> Originally known as The Landlord's Game, this proto-Monopoly had a number of familiar features—the continuous rectangular path, the Go to Jail corner—but it appeared too complex and didactic to have mass appeal. (Ferguson 2008, 230–231)

Here we have a developmental process that passes through The Landlord's Game on the way to Monopoly. In addition, the Landlord's Game resembles and prefigures Monopoly. It thereby qualifies as a proto-Monopoly.

In *The Information*, James Gleick gives a sketch of the life of information theorist Claude Shannon, which includes the following sentence:

> As a first-year research assistant at MIT, he worked on a hundred-ton proto-computer, Vannevar Bush's Differential Analyzer, which could solve equations with great rotating gears, shafts, and wheels. (Gleick 2011, 6)

Here we have a developmental process that passes through Vannevar Bush's Differential Analyzer (and other such clunky devices) on the way to today's high-speed electronic computers. In addition, Vannevar Bush's Differential Analyzer resembles and prefigures today's high-speed computers. It thereby qualifies as a proto-computer.

In *The Emperor of All Maladies*, Siddhartha Mukherjee discusses the role of "proto-oncogenes" in carcinogenesis:

> The crucial implication of the Varmus and Bishop experiment was that a precursor of a cancer-causing gene—the "proto-oncogene," as Bishop and Varmus called it—was a normal cellular gene. Mutations induced by chemicals or X-rays caused cancer not by "inserting" foreign genes into cells, but by activating such endogenous proto-oncogenes. (Mukherjee 2010, 362)

Here we have a developmental process that converts a normal gene that regulates the normal process of cell division into a gene that facilitates cancerous cell division. In addition, the normal gene resembles and prefigures the gene that facilitates cancerous cell division. It thereby qualifies as a proto-oncogene.

<>

Gregg Rosenberg is a panexperientialist whose view I discussed briefly in chapter 12. In Rosenberg's terminology, consciousness requires a certain kind of cognitive setting, which the simple and alien experiential properties that he attributes to electrons do not have. He thus says that the experience that electrons have is protoconscious but not conscious. This is all pretty vague. Rosenberg does not make it clear how he draws the line between noncognitive, protoconscious experience and cognitive, conscious experience. Still, his use of "protoconscious" clearly fits the pattern I have been discussing. He imagines a developmental process that passes from noncognitive experience to cognitive and thus conscious experience. The noncognitive experience resembles and prefigures the conscious experience by virtue of being experiential. The noncognitive experience thereby qualifies as protoconscious.

Now we come to the ambiguities.

David Chalmers has a classification system for views on the mind-body problem that includes the category of "type-F monism." The views that Chalmers places in this category feature a distinction similar to Bertrand Russell's distinction, which I criticized in chapter 17, between "structure" and/or "mathematical properties" that can be inferred from perception and "intrinsic character" or "intrinsic quality" that cannot be inferred from perception. Instead of Russell's terms "intrinsic character" and "intrinsic quality," Chalmers uses the term "intrinsic properties." He says that there is a form of type-F monism according to which some "intrinsic properties" are "protophenomenal properties":

> Perhaps the intrinsic properties of the physical world are themselves phenomenal properties. Or perhaps the intrinsic properties of the physical world are not phenomenal properties but nevertheless constitute phenomenal properties: that is, perhaps they are protophenomenal properties. (Chalmers 2010, 133)

Rosenberg compares his use of "protoconscious" with Chalmers's use of "protophenomenal" in the following passage, which accords with my understanding of Chalmers:

> The properties of protoconsciousness can be usefully and explicitly contrasted with the *protophenomenal* properties proposed in Chalmers (1996). According to Chalmers, protophenomenal properties would be fundamental nonexperiential, nonphenomenal properties. By hypothesis, in proper combination protophenomenal properties could become experienced phenomenal properties. Chalmers leaves open what contexts can provide the proper combinations, but we can presume only cognitive contexts work because the proposal seems designed to avoid panexperientialism. In contrast with protophenomenal properties, the properties of protoconsciousness are experiential

properties properly considered phenomenal, but they do not require an associated cognitive engine to be experienced. (Rosenberg 2004, 96–97; his emphasis)

The idea that experiential ("phenomenal") intrinsic properties are constituted by certain combinations of nonexperiential intrinsic properties is an idea that type-F monism as defined by Chalmers adds to Russell's view. According to Russell, there are swarms of ephemerons that have both structure and intrinsic character. Additionally, Russell says that some intrinsic character is experiential and some is not experiential, but to my knowledge he does not hypothesize any particular relation between the two.

Chalmers notes that his description of type-F monism is missing important details:

> ... we need a much better understanding of the *compositional* principles of phenomenology; that is, the principles by which phenomenal properties can be composed or constituted from underlying phenomenal properties, or protophenomenal properties. (Chalmers 2010, 136; his emphasis)

More fundamentally, I question whether this description makes sense. I don't know what it means to say that one property is constituted from other properties. I understand what it is for an object to be constituted from other objects, as a house is constituted from boards and bricks and such. I understand what it is for an object to be constituted from a set of ingredients that get mixed together, as a cake is constituted from flour and eggs and such. But I don't think of properties in this way. This issue seems somewhat ill-defined, however, so I will not pursue it further.

The key point to notice about Chalmers's use of the prefix "proto" in "protophenomenal" is that it does not fit the three-condition pattern that I described at the beginning of this chapter. It does not fit that pattern because his description of type-F monism does not include any claim that protophenomenal properties *resemble* phenomenal properties in some way. Recall from chapter 17 that Russell hints repeatedly at a resemblance between experiential intrinsic character and nonexperiential intrinsic character, but fails to specify any resemblance or to explain how there could be a resemblance between experience and something nonexperiential. Chalmers does not have this problem because he does not say that there is any resemblance between phenomenal properties and protophenomenal properties. He is willing to call a property protophenomenal based solely on its having a constitutional relation to a phenomenal property, whether or not it has any resemblance to the phenomenal property that it helps to constitute.

In his essay "Panpsychism" Thomas Nagel assesses the plausibility of a certain hypothesis that involves "proto-mental properties" (hyphenated). Nagel describes these as "properties of matter ... discoverable by explanatory *inference* from observable mental phenomena" (Nagel 1979a, 184; my emphasis) and "properties that *imply* the appearance of different mental phenomena when the matter is combined in different ways" (Nagel 1979a, 182; my emphasis). After making a case that such properties exist, he proceeds to cast doubt on the idea as follows:

> ... it is difficult to imagine how a chain of explanatory inference could ever get from mental states of whole animals back to the proto-mental properties of dead matter. It is a kind of breakdown we cannot envision; perhaps it is unintelligible. (Nagel 1979a, 194)

There are three significant points of comparison—one difference and two similarities—between the notion of "proto-mental properties" in Nagel's essay and the notion of "protophenomenal properties" in the writing of Chalmers. The difference is that Nagel and Chalmers imagine different relations connecting proto-x to x. Nagel imagines a cognitive relation of inference or implication, whereas Chalmers imagines a noncognitive relation of constitution or composition. These two relations are logically compatible with one another, but either could obtain without the other.

One similarity between Nagel and Chalmers is that neither of them says anything about resemblance between proto-x and x. Like Chalmers's use of "protophenomenal," Nagel's use of "proto-mental" does not fit the three-condition pattern that I described at the beginning of this chapter because it does not include the resemblance condition. Another similarity between Nagel and Chalmers is that both of them think of their "proto" properties as fundamental properties. In this respect too, Nagel and Chalmers deviate from the three-condition pattern. There is no suggestion that proto-wrists, proto-Monopoly, proto-computers, or proto-oncogenes are in any sense fundamental; all of these are highly developed precursors of other highly developed things that come after them.

I question the appropriateness of the term "panpsychism" for the view that Nagel describes in this essay. To me, the core idea of panpsychism is precisely that there is a significant resemblance between familiar mental properties and certain properties of electrons and other fundamental particles. Given this understanding of panpsychism, if resemblance between mental properties and "proto-mental properties" is not part of the view Nagel is discussing, "panpsychism" is not an appropriate term for it. Perhaps the view described in Nagel's essay is in some sense a conceptual neighbor of panpsychism, but it differs from panpsychism as ordinarily understood in an important way.

In his book *Mind and Cosmos*, Nagel assesses the plausibility of a variety of other hypotheses concerning the foundations of experience. His descriptions of some of these hypotheses include the terms "protomental" (not hyphenated) and "protopsychic," which he seems to use interchangeably. In the following passage, he seems to associate the prefix "proto" with a compositional notion very similar to that of Chalmers:

> The space-time framework of the physical world makes the physical part-whole relation immediately graspable, geometrically, but we have no comparably clear idea of a part-whole relation for mental reality—no idea how mental states at the level of organisms could be composed out of the properties of microelements, whether those properties are similar in type to our experiential properties or different. Yet a mentalistic reductionism would presumably have to find the protomental parts in a monist counterpart of the physical parts of the organism, and would have to include a theory of how they combine into conscious wholes. (Nagel 2012, 62–63)

Alluding to this passage a bit further along, he refers to "the mind's protomental parts" (Nagel 2012, 88) and acknowledges "serious problems about the mental part-whole relationship" (Nagel 2012, 87). In addition, it may be that part of what Nagel means by "proto" in this book is a cognitive relation of inference and/or implication such as that described in his essay "Panpsychism," because he envisions protomental or protopsychic properties playing a role in a certain sort of explanation:

> A naturalistic expansion of evolutionary theory to account for consciousness would not refer to the intentions of a

> designer. But if it aspires to explain the appearance of consciousness *as such*, it would have to offer some account of why the appearance of conscious organisms, and not merely of behaviorally complex organisms, was likely. (Nagel 2012, 48; his emphasis)

> To explain consciousness, a physical evolutionary history would have to show why it was likely that organisms *of the kind that have consciousness* would arise. (Nagel 2012, 60; his emphasis)

I note in passing that the goal of explaining why the evolution of conscious animals "was likely" is an unusual one. The usual aspiration is to explain *how* consciousness comes to be, not why its development was likely. Further, I do not know what Nagel means in saying that the evolution of conscious animals "was likely," or what reason there is to believe that this "was likely." Setting these puzzles aside, the important point here is that Nagel's *Mind and Cosmos* is like his essay "Panpsychism" and the work of Chalmers in using the prefix "proto" in a way that does not fit the three-condition pattern that I laid out at the beginning of this chapter. The usages of Nagel and Chalmers do not satisfy the condition that proto-x resembles x. At the same time, they bring in various additional conditions, such as that proto-x helps to constitute x, proto-x is inferentially related to x, and proto-x is fundamental.

Unlike Chalmers and Nagel, I do not see any reason to think that there are fundamental properties that can be combined to *constitute* experiential properties, or fundamental properties that can be *inferred from* experiential properties. Accordingly, there is no place for the prefix "proto" in Chalmers's sense or Nagel's sense in a description of the quadrants view. On the other hand, using the prefix "proto" in the three-condition sense that allows a certain type of fish to have a proto-wrist, I could say that the quadrants view is compatible with the possibility that the evolutionary dawn of experiential properties followed upon an epoch of protoexperiential properties that dawned a while earlier. In chapter 7, I alluded to this possibility using the word "quasi-experiential." I do not use the prefix "proto" in this way, however, because doing so would invite confusion with the views of Chalmers and Nagel while adding nothing to what I find myself able to say about the quadrants view without the prefix "proto." As with so many other terms discussed in this book, the ambiguity of the prefix "proto" gives me a good reason to avoid it.

20

Identity and Existence

According to the quadrants view, there is no overlap between outwardly observable properties and inwardly observable experiential properties. Experiential properties are never outwardly observable because they have no evidentiary effects (and perhaps no effects at all) on the spatial transmissions that outward observation depends on. This basic tenet of the quadrants view conflicts with some so-called identity theories. In this chapter and the next I explain what is wrong with identity theories in general, and with identity theories that conflict with the quadrants view in particular.

Identity theories involve identity statements. In general, though, proponents of identity theories do not put forward actual identity statements. Rather, they argue that it is reasonable to anticipate the future formulation of certain problem-solving identity statements after more brain research has been done. The anticipated identity statements are of a certain type, which can be described as *true two-term identity statements about the real world*. Following are three frequently cited examples of this type of identity statement:

Mark Twain is Samuel Clemens.

The morning star is the evening star.

Water is H_2O.

Such statements differ in an obvious way from one-term identity statements, which use the same term twice:

Mark Twain is Mark Twain.

The morning star is the morning star.

Water is water.

They also differ from two-term identity statements that are about a work of fiction:

Superman is Clark Kent.

Dr. Jekyll is Mr. Hyde.

Finally, they differ from false two-term identity statements about the real world, such as the following:

Mark Twain is Abraham Lincoln.

The morning star is Alpha Centauri.

Water is CO_2.

A crucial difference between a true two-term identity statement about the real world and a false two-term identity statement about the real world is the number of referents involved. Behind a true two-term identity statement there is a single referent, which both terms refer to. Behind a false two-term identity statement there are two referents, each referred to by one of the terms.

Some writers cite statements such as the following as examples of true identity statements:

Heat is the motion of molecules.

Lightning is an electrical discharge.

These are true statements, I believe, but they are not true identity statements, because the phrase after the "is" has a broader extent than the phrase before the "is." "The motion of molecules" includes not only heat, but also wind, molecular diffusion, circulating blood, and much more. "An electrical discharge" includes not only lightning, but also sparks in an internal combustion engine, laboratory demonstrations of discharging capacitors, and the little shocks that you can get after shuffling across a carpet on a winter day. I think this type of mistake is typically the result of omitting important qualifications. For example, one can produce the statement "Lightning is an electrical discharge" by abridging longer statements such as the following:

> ... lightning is identical with a sudden large-scale discharge of electrons between clouds, or between the atmosphere and the ground. (Churchland 1988, 26)

> ... lightning is ... the massive, sudden discharge of the collective electrical charge generated by the movement of many slightly charged water droplets or ice crystals that form the clouds. (Guttenplan 1994, 91)

These longer statements are, at least arguably, true identity statements about the real world.

<>

The central claim of this chapter is that anything that can be truly said using a two-term identity statement can also be said using other sentence forms. This may not sound too surprising, given the resources and versatility of human language, but it has an important implication. What exactly do the many users of the phrase "identity theory" mean by it? I have never seen a discussion of this phrase, but it is easy to get the impression that many people think of an identity theory as a theory in which two-term identity statements play *an essential and indispensable role*, a theory that can *only* be stated using two-term identity statements. Would people call a theory an "identity theory" just because two-term identity statements provide one way to state it? Would people devote so much attention to two-term identity statements if they thought that such statements were merely one way to state a theory that can also be stated in other ways? The trouble is that *there is no such thing as an identity theory* in this sense if anything that can be truly said using a two-term identity statement can also be said using other sentence forms. There are theories that can be expressed wholly or partly by means of two-term identity statements, but there are no theories that *require* two-term identity statements for their expression. This fact deprives two-term identity statements of the importance that they seem to

279

have in the eyes of some. It suggests that the extensive, decades-long discussion of two-term identity statements in connection with the mind-body problem amounts to a misguided fashion that has been sustained by a misconception. It also opens up a useful critical perspective on so-called identity theories, which I employ in the next chapter.

Suppose that someone publishes a new theory entirely in italics, perhaps because he is so excited about it, and suppose that he calls it an italics theory. It will not be long before readers point out that the content of this theory would not change if the whole thing was printed in another font. The use of italics is incidental, and therefore the use of the phrase "italics theory" is misconceived and misleading. The situation is the same with so-called identity theories, only less obviously so.

Imagine a strict ban on the use of two-term identity statements. Perhaps an oddball dictator imposes the death penalty on anyone suspected of using one. The following argument shows that such a ban would not reduce anyone's ability to describe the world.

In the TV game show Jeopardy, contestants must state every answer in the form of a question. This is a significant constraint, but we can be confident that it will never prevent a contestant from giving the correct answer because there is a well understood procedure for putting the content of any answer in the grammatical form of a question. Analogously, suppose that you set yourself the task of describing a certain part or aspect of the world, subject to the constraint that you must refer to nothing in more than one way. For example, if you are writing about American literature and you refer to a certain author as "Mark Twain," then you cannot also refer to him as "Samuel Clemens" or "the author of *Tom Sawyer*" or even as "he." Call this the single-path-of-reference constraint. This is a significant constraint, but you can be confident that it does not prevent you from making any points because as long as you have one way to refer to something, you can use that way as often as necessary to say whatever needs to be said about that referent. For example, using the single path of reference "Mark Twain," you can write the following:

> Mark Twain's given name was "Samuel Clemens." Mark Twain wrote *Tom Sawyer*, *Huckleberry Finn*, and many other books. Mark Twain was born in Missouri. Mark Twain took a special interest in the Mississippi River ...

Of course, such repetition of a referring expression makes for monotonous prose, but that is beside the point. The point is that any true account can be rewritten subject to the single-path-of-reference constraint without loss of content.

To this point add the fact that every true two-term identity statement violates the single-path-of-reference constraint. Such a statement contains two different terms that have a common referent; that is a plain violation. Since it is never necessary to violate the single-path-of-reference constraint, it is never necessary to use a two-term identity statement. Whatever a two-term identity statement says about the world, there is always a way to say it using another sentence form.

I think this is an excellent argument, but you might find it a bit too abstract to be convincing. It would be nice to supplement it with an explicit procedure that takes as input any true two-term identity statement and gives as output a true statement of another form that says everything that the two-term identity statement says. In a moment I will specify such a procedure, which gives as output what I call a companion existence statement. Before I do this, however, I need to discuss a certain issue concerning the interpretation of two-term identity statements.

<>

Many writers maintain that part of the content of every two-term identity statement is that the common referent of the two terms satisfies the identity relation, or, in other words, that the common referent of the two terms is identical to itself. Following are two reasons to question this claim.

First, many two-term identity statements seem to be mere stylistic variants of statements that are not identity statements. I will give three pairs of examples.

The following two sentences are different ways of stating the same fact of authorship:

Mark Twain is the author of *Tom Sawyer*.

Mark Twain wrote *Tom Sawyer*.

It seems odd to hold that the "is" statement makes an additional point that the "wrote" statement does not make, namely that the man in question is identical to himself.

The following two sentences are different ways of making a certain point about preferences:

Chocolate is my favorite flavor.

I prefer chocolate to all other flavors.

It seems odd to hold that the "is" statement makes an additional point that the "prefer" statement does not make, namely that the flavor in question is identical to itself.

The following two sentences are different ways of making a certain point about the relationship between a particular river and a particular lake:

The source of the Mississippi River is Lake Itasca.

The Mississippi River originates in Lake Itasca.

It seems odd to hold that the "is" statement makes an additional point that the "originates" statement does not make, namely that the lake in question is identical to itself.

Of course, in each of the preceding "is" statements, the common referent of the two referring terms is identical to itself. I don't dispute that. My claim is that the self-identity of the common referent does not seem to be part of what these "is" statements assert.

The second reason is that it is a generally acknowledged truism that everything that exists is identical to itself. We all know this, and we all know that we all know it. So why would someone bother to point out that some particular thing is identical to itself? It goes without saying. Consider the following analogy. You make a new acquaintance. She shares a few facts about herself—her name, her occupation, where she went to school. Then she says, "And I breathe air." You would find this very odd, because it's a generally acknowledged fact that every living person breathes air. We all know this, and we all know that we all know it. People don't bother to state this fact because it goes without saying. The idea that there's a sentence form that people use to say that one or another particular thing is identical to itself seems odd for the same reason.

Against this, someone might say that there is something odd, if not downright contradictory, about the idea that an identity statement does not say that something is identical to itself. After all, it's an identity statement. However, this argument is circular. These "identity statements" were labeled as such by people who took them to make claims of self-identity. This has been the usual way to read them in certain circles, but it is not the only reasonable way to read them, or even the most reasonable way to read them, as the discussion here shows. Suppose we instead call these statements co-reference statements, reflecting the fact that they contain two terms that have a common referent. There is no contradiction in the claim that a co-reference statement does not say that the common referent of its two terms is identical to itself. Or suppose we dispense with labels and consider sentences in their individual particularity. There is no contradiction in the claim that the statement "Mark Twain is Samuel Clemens" does not say that the man in question is identical to himself.

The upshot is that two-term identity statements or co-reference statements have a systematic ambiguity. They can be understood either as including or as not including claims of self-identity. Accordingly, I will give due consideration to both ways of understanding these statements.

<>

The general procedure for constructing a companion existence statement from a true two-term identity statement has four steps, which are laid out below. The optional step 3 is included because the procedure must work for both ways of understanding two-term identity statements. To accommodate the way of understanding these statements that includes a claim of self-identity, perform step 3. To accommodate the way of understanding these statements that does not include a claim of self-identity, omit step 3.

> *Step 1.* Begin with the words "There is a"
>
> (Use another form of the verb "to be" if appropriate.)
>
> *Step 2.* Add a word that designates a category to which the common referent of the two-terms of the identity statement belongs.
>
> *Step 3 (optional).* Add the words "that is identical to itself and"

(If the common referent is an individual human being, use "himself" or "herself" instead of "itself," as appropriate.)

Step 4. Add "that" (or "who" in the case of an individual human being) followed by a description of conditions that includes all the conditions that a thing must satisfy in order to be the common referent of both terms of the identity statement. You can include other conditions that are not associated with the terms of the identity statement. The only constraint on what you can include is that the resulting statement must be true.

To illustrate how this procedure works, I have constructed a pair of companion existence statements for each of the three true two-term identity statements about the real world that I cited at the beginning of this chapter. In each companion existence statement, the category term that is added in step 2 is in italics and the contribution of the optional step 3 is in parentheses. From "Mark Twain is Samuel Clemens" one can construct the following companion existence statements:

There was a *man* (who was identical to himself and) who was known to the public as Mark Twain but whose given name was Samuel Clemens.

There was an *American author* (who was identical to himself and) who wrote books under the pseudonym "Mark Twain" and whose given name was Samuel Clemens.

From "The morning star is the evening star" one can construct the following companion existence statements:

There is a *planet* (that is identical to itself and) that can be seen in the eastern sky before sunrise, where it is known as the morning star, and that can also be seen in the western sky after sunset, where it is known as the evening star.

There is a *celestial body* (that is identical to itself and) that became known as the morning star because it looks like a bright star in the eastern sky before sunrise, and that also became known as the evening star because it looks like a bright star in the western sky after sunset.

From "Water is H$_2$O" one can construct the following companion existence statements:

> There is a *type of molecule* (that is identical to itself and) that contains one oxygen nucleus and two hydrogen nuclei and that, in huge numbers, constitutes the liquid that fills lakes and rivers.

> There is a *molecular species* (that is identical to itself and) that is formed by the combination of two hydrogen atoms with one oxygen atom and that, in huge numbers, constitutes the liquid that comes out of our household faucets.

Note the following important points about this procedure:

1. Whatever the two terms of a two-term identity statement are, you can transfer their content to the description of conditions that you write in step 4. This is simply a matter of combining the content of the two terms into a single description. Therefore, the procedure works for every two-term identity statement.

2. Many choices are available for the category term that you add in step 2. These choices include broad and fuzzy category terms such as "thing," "entity," and "phenomenon." However, the more specific and definite the category term is, the more informative the resulting companion existence statement will be.

3. Steps 2 and 4 of the procedure enable you to add descriptive content that is not in the identity statement. Therefore, a companion existence statement is generally not equivalent to the identity statement from which it is made. It generally says more.

4. Steps 2 and 4 give you choices. Therefore, the procedure does not yield a unique companion existence statement for each identity statement. Rather, for a given identity statement, you can use this procedure to construct an indefinite number of companion existence statements. They will all be rather similar, though, because they all share a core

of content that is dictated by the identity statement from which they are made.

5. Step 4 enables you to add an indefinite amount of information about the common referent of the two terms of the identity statement. Therefore, a companion existence statement can get quite long. If it becomes too long, you can break it up into two or more sentences. There is no need to pack everything into a single grammatical sentence.

6. The only thing that depends on the difference between the two ways of understanding two-term identity statements is whether the parenthesized output of step 3 is included. Since none of the preceding points concerns step 3, they all apply regardless of which way you understand two-term identity statements.

The fact that this procedure is completely general (point 1) confirms the conclusion that there is no such thing as a theory that can be stated only by means of two-term identity statements. Rather, two-term identity statements provide a way of stating certain theories that can also be stated with existence statements. A theorist who is prevented from using two-term identity statements is not thereby limited in the theories he can express. He can make claims about what things exist, what characteristics these things have, and how they are related to other things, including the words and phrases of any language. He can even state that each existing thing is identical to itself, if he considers that point worth making. Like the Jeopardy rule that all answers must be given in the form of a question, a rule prohibiting the use of two-term identity statements does not prevent the communication of any cognitive content.

One can go a step further. The examples show that if you understand two-term identity statements as including claims of self-identity, then all claims of self-identity are covered by the same stock clause in the companion existence statements. Moreover, a companion existence statement that includes this self-identity clause is true if and only if the statement without this clause is true. The substantive issue is whether or not something exists that satisfies a certain set of conditions. If such a thing exists, then of course it is identical to itself. If no such thing exists, then the question of self-identity does not arise. Therefore, two-term identity statements are a way to state certain theories that can also be stated by means of existence statements that include no self-identity clause. The self-identity claim is implicit in the existence claim. It adds nothing of substance to the theory.

There are situations in which a true two-term identity statement can disabuse someone of a false belief in the existence of a certain pair of things. For someone who is under the impression that the morning star and the evening star are different objects, the statement "The morning star is the evening star" is informative in a surprising way. For the slow-to-catch-on characters in *The Adventures of Superman*, the statement "Superman is Clark Kent" would carry a similar payload of revelation. I believe this is the characteristic of two-term identity statements that has recommended them to some writers. The idea was to use two-term identity statements in an analogous manner in order to disabuse people of their false belief in the twoness of inwardly observable properties and outwardly observable properties.

But a two-term identity statement is merely one way to correct this sort of false impression; it is not the only way. A companion existence statement can convey the surprising revelation just as well, if not better. Instead of saying "The morning star is the evening star," one can say "There is a single planet that gives rise to both that bright star appearance shortly before sunrise and that bright star appearance shortly after sunset." Instead of saying "Superman is Clark Kent," one can say "There is a single individual who calls himself Clark Kent when working as a journalist and Superman when dressed in a colorful cape and using his superhuman powers to apprehend criminals." Likewise, if inwardly observable properties are not distinct from outwardly observable properties, one can make that point using companion existence statements instead of two-term identity statements. This again confirms the main claim of this chapter: whatever can be said with two-term identity statements can be said without them, by using companion existence statements.

<>

Saul Kripke (1971, 1980) has argued at length that some two-term identity statements express what he calls "necessary a posteriori truths." One example is "Water is H_2O." I think this claim is misleading in two respects.

First, a two-term identity statement does not have any cognitive content that is both necessary and a posteriori. What is a posteriori is the belief that something satisfying a certain set of conditions exists. In addition, there is a necessary truth that whatever exists is identical to itself. From these two truths one can deduce a third, as in the following example:

Water-H_2O exists. (A posteriori premise)

Whatever exists is identical to itself. (Necessary premise)

Therefore, water-H_2O is identical to itself. (Deduced from this pair of premises)

One can regard a two-term identity statement as a cognitive package that expresses both the a posteriori premise of such an argument and the necessary-truth-dependent conclusion. But calling this package a "necessary a posteriori truth" is like calling an appetizer on a toothpick an edible wooden object. Part of this item is edible and part of it is wooden, but it contains no edible wood. Likewise, a two-term identity statement expresses no a posteriori necessity.

Second, Kripke's whole discussion presupposes the way of understanding two-term identity statements that includes a claim of self-identity. I have shown that it is reasonable, maybe even more reasonable, to understand two-term identity statements as not claiming that the common referent of the two terms is identical to itself. If one understands two-term identity statements in this way, there is nothing necessary about them.

21

Six Identity Theorists

This chapter surveys the thinking of six identity theorists—Donald Davidson, J. J. C. Smart, Herbert Feigl, David Papineau, Christopher Hill, and Paul Churchland. It discusses details of these six views, illustrating in the process two broad themes.

One of these themes is indefiniteness and lack of clarity. Identity theorists often content themselves with ill-defined referring terms that make it difficult to form a clear idea of what the terms are supposed to refer to. The true two-term identity statements that I cited as examples at the start of the previous chapter have terms that refer with a high degree of clarity to a person, a celestial body, and a type of molecule, respectively. Such clarity is invariably absent in the writings of those who propose identity theories that bear on the mind-body problem. A good way to demonstrate this lack of clarity is to try to construct companion existence statements that focus attention on the nature of the common referent of the two purportedly co-referring terms.

Referring is a verbal gesture that can be done with varying degrees of precision. People often use fuzzy referring terms, and in many contexts there is nothing wrong with that. For example, referring terms such as "yesterday's weather" or "the global economy" can serve a purpose despite their fuzziness. This is not the case, however, when a referring term is part of a two-term identity statement, because the truth or falsehood of such a statement depends on whether or not its two terms are co-referential. How can one decide whether two terms are co-referential if one can't form a clear idea of what either term refers to? Even one fuzzy term makes a mess of a co-reference claim. There is a poem by Jean Garrigue with the title "Why the Heart Has Dreams Is Why the Mind Goes Mad." This is a two-term identity statement whose referring terms are extremely ill-defined. As a consequence, it has no determinate truth value.

That's fine for poetry, but not for a theory that is meant to be literally true. For some reason, many identity theorists seem not to appreciate the exacting nature of the sentence form that they consider so important. The referring terms that they propose are not *intentionally* fuzzy, like the terms in the title of this poem, but many are far too fuzzy to yield a two-term identity statement that one can feel one understands. To make matters even worse, identity theorists often do not specify actual referring terms, but merely indicate the general character of the referring terms that their theories need. All in all, the fog in this region is rather thick.

The second theme is variety. The writings of identity theorists are definite enough to make it clear that they do not all imagine the same sort of thing playing the role of common referent. The following table shows the categories to which the identity theorists discussed in this chapter assign the common referents of their identity statements:

Identity Theorist	Category That Referents Belong To
Donald Davidson	events
J. J. C. Smart	brain processes
Herbert Feigl	raw feels
David Papineau	physical properties
Christopher Hill	brain processes, brain states
Paul Churchland	sensory coding vectors

There is presumably a certain amount of overlap among these categories, but there are also significant differences. It is therefore a mistake to speak of *the* mind-body identity theory, as some commentators do. Various identity theories, differing from each other in various ways, have been proposed.

The underlying cause of this variety, I believe, is that there is nothing that can successfully play the role of common referent. Identity theorists have groped around in search of something that doesn't exist, and as a result different theorists have latched on to different things.

One consequence of this variety is that some identity theories conflict with the quadrants view and some do not. All identity theories have problems, but I am mainly concerned to point out the problems of identity theories that conflict with the quadrants view.

<>

Donald Davidson imagines identity statements whose two terms refer to a single "event." One term refers to the "event" by way of "mental" vocabulary and the other term refers to it by way of "physical" vocabulary. Unlike Bertrand Russell, as described in chapter 16, Davidson does not specify a nonstandard meaning for the word "event"; he seems to regard himself as using this word in its everyday sense. As I noted in my discussion of Russell, "events" in the everyday sense are interest-relative human inventions, not naturally bounded parts of the world. The world does not contain naturally bounded events in the way that it contains naturally bounded people, planets, and molecules. Accordingly, there are no well-defined events to serve as the common referents of the two terms of a two-term identity statement. Davidson seems not to realize that the word "event" has this loose, interest-relative character. He writes as if the word "event" in ordinary English works in the same way as "peanut" and "pumpkin," designating a class of naturally bounded something-or-others. He lists various subclasses of "events" that he presumes to be well-defined:

> perceivings, rememberings, decisions, and actions (Davidson 2001b, 207)

> perceivings, notings, calculations, judgements, decisions, intentional actions, and changes of belief (Davidson 2001b, 208)

> perceivings, rememberings, the acquisition and loss of knowledge, and intentional actions (Davidson 2001c, 231)

He does not give many examples of expressions that refer to a particular "event" using the "mental" vocabulary, but here are two:

> remembering that one has left a zipper open (Davidson 2001a, 176)

> deciding to schuss the headwall (Davidson 2001a, 176)

As far as I can tell, he does not give any examples of expressions that refer to such an "event" using the "physical" vocabulary. Presumably his thought is that these expressions will emerge one day from brain research.

What sort of thing could the two referring terms in such an identity

291

statement refer to? A companion existence statement might start "There is a remembering..." or "There is a deciding..." But what does a remembering or a deciding consist of? Does "the remembering" that is the referent of "remembering that one has left a zipper open" include only the inwardly observable features of a certain recollection experience? Does it also include connections between this recollection experience and unconsciously stored memory content? Does it perhaps include all the properties of all the electrons, protons, and neutrons that contribute in any way to the occurrence of the recollection experience? Does "the deciding" that is the referent of "deciding to schuss the headwall" include only the final moment of decision? Does it also include all the vacillation and deliberation leading up to the final moment of decision as well as the unconsciously stored memory content that plays a role in the decision process? Does it perhaps include all the properties of all the electrons, protons, and neutrons that contribute in any way to the occurrence of the decision experience? Such questions do not have correct answers, because there are no boundaries in nature to give them correct answers. Inwardly one encounters a seamless "stream of consciousness" while outwardly one encounters a seamless stream of metabolism and electrochemistry. The life of a person is a seamless ongoing totality, not a mosaic of discrete "events."

Although I believe Davidson's identity theory is defective in the way I have just described, showing this is not integral to a defense of the quadrants view because Davidson's identity theory does not conflict with the quadrants view. Davidson puts no limit on the complexity of the "events" that he speaks of. He assumes that each of these "events" has properties that support a "mental" referring expression and other properties that support a "physical" referring expression. Without endorsing a view of this sort, Jerome Shaffer spells out the envisioned situation in the following passage, which amounts to a kind of collective companion existence statement for Davidson-style identity statements:

> That is to say, there will be this class of events which will be known to occur either on the basis of neurological observations or by introspection. There will still be privileged access to these events, but there will also be public access to them. The event I know to have occurred on the basis of introspection will turn out to be one and the same as the event you know to have occurred by neurological observation. This is possible because one and the same event will have both physical and non-physical features. (Shaffer 1970, 139)

According to the quadrants view, all electronuclear objects have outwardly observable properties and remote properties, and many living electronuclear objects also have experiential properties. Davidson is not wrong in thinking that there are things that have both inwardly observable "mental" properties and outwardly observable "physical" properties. His mistake is to think that there are things having properties of these two types that are much smaller than a whole living animal, naturally bounded, and associated with the word "event" in its everyday sense.

This point can be generalized. The quadrants view is logically compatible with the existence of any complex entity that has some properties that are outwardly observable and some properties that are not outwardly observable. Consequently, no two-term identity statement whose terms refer to something having this kind of complexity poses a challenge to the quadrants view. The challenge to the quadrants view comes from identity statements whose two terms are said to refer to something all of whose properties are outwardly observable. The other five identity theorists discussed in this chapter envision identity statements of this sort.

<>

J. J. C. Smart envisions two-term identity statements whose terms refer to "brain processes," which he thinks of as having only outwardly observable "neurological properties" (Smart 1959, 61). In each identity statement, one of the terms uses a description of the neurological properties of the brain process to refer to it and the other term uses our everyday vocabulary for talking about sensations. As an example of everyday sensation vocabulary, Smart uses the following:

> Suppose that I report that I have at this moment a roundish,
> blurry-edged after-image which is yellow towards its edge and
> is orange towards its center. (Smart 1959, 52)

According to Smart's theory, the relation between a verbal sensation report and what it is a report of is one of causal dependence without comprehension, as when a computer "reports" an error condition:

> The strength of my reply depends on the possibility of our
> being able to report that one thing is like another without
> being able to state the respect in which it is like. I do not see

> why this should not be so. If we think cybernetically about the nervous system we can envisage it as able to respond to certain likenesses of its internal processes without being able to do more. (Smart 1959, 61)

> For this account to be successful, it is necessary that we should be able to report two processes as like one another without being able to say in what respect they are alike. An experience of having an after-image may be classified as like the experience I have when I see an orange, and this likeness, on my view, must consist in a similarity of neuro-physiological pattern. But of course we are not immediately aware of the pattern; at most we are able to report the similarity. (Smart 1963b, 95)

This account acknowledges the obvious fact that people have and use a large sensation vocabulary, but it denies the existence of any properties of human beings that are inwardly observable.

Smart gives no examples of actual identity statements. Here is a sketch of one that uses his example of a sensation report:

> My current sensation of a roundish, blurry-edged after-image which is yellow towards its edge and orange towards its center is a brain process that has neurological properties X, Y, and Z and no inwardly observable properties.

Here is a companion existence statement:

> There is a brain process that has neurological properties X, Y, and Z and no inwardly observable properties, which the person in whom it occurs refers to as a sensation of a roundish, blurry-edged after-image which is yellowish towards its edge and orange towards its center.

There are three major problems here.

First, like Davidson's "events," Smart's "brain processes" lack natural definition. Smart says that a brain process is "a very complex process involving vast numbers of neurons" (Smart 1963a, 165). But which neurons, which molecules and ions within those neurons, and which properties of those neurons, molecules, and ions are included in a particular "brain process"? How many "brain processes" are going on in me right now? How many "brain processes" have come to an end in me since I ate breakfast? These questions do not have correct answers, because they presuppose a false conception of the brain. A living brain is a hive of activity. It seems harmless to say, speaking loosely, that this activity consists of "brain processes," but in fact there are no discrete, naturally bounded brain-process-entities that can serve as the common referent of the two referring terms of a two-term identity statement.

Second, Smart's identity theory involves a magical exception to the normal rules of reference. Normally, a descriptive expression succeeds in referring to something by specifying conditions that the referent satisfies. How then does a description such as "a sensation of a roundish, blurry-edged after-image which is yellowish towards its edge and orange towards its center" manage to refer to a "brain process" that has no inwardly observable experiential properties? Such a purely outward "brain process" would not satisfy such a description, and so could not be referred to by it, if normal rules of reference are in force.

Third, Smart's identity theory depends on the general claim that human beings are not acquainted with their sensations unless they happen to be brain scientists making outward observations of their own brains. This claim defies credulity. I might not be able to learn *much* about myself through inward observation, but inward observation is informative in its way. As further support for this point, note that there are various judgments that people routinely make about their sensations, such as whether they are pleasant or unpleasant, familiar or novel, worth paying to obtain, or worth paying to get rid of. Such judgments make no sense unless they are based on acquaintance with the sensations being judged. If Smart really found it plausible to deny the possibility of inward-observation-based acquaintance with sensations, I suspect this was due to his focus on *reporting* sensations as opposed to simply *having* sensations. If you set aside the business of communicating with other people and simply attend to your current sensations or to recollections of your past sensations, I think you will find the conclusion inescapable that inward observation can tell you something about yourself.

<>

Herbert Feigl envisions two-term identity statements whose terms refer to "raw feels." He names various categories of "raw feels" that he thinks fit to play this role:

> aches, pains, tickles, moods, emotions, etc. (Feigl 1967, 26)

> directly experienced sensations, thoughts, feelings, emotions, etc. (Feigl 1967, 79)

Feigl describes the identity statements that he envisions in passages such as the following:

> The identity thesis which I wish to clarify and to defend asserts that the states of direct experience which conscious human beings "live through," and those which we confidently ascribe to some of the higher animals, are identical with certain (presumably configurational) aspects of the neural processes in those organisms. (Feigl 1967, 79)

> The "mental" states or events (in the sense of raw feels) are the referents (the denotata) of the phenomenal terms of the language of introspection, as well as of certain terms of the neurophysiological language. (Feigl 1967, 80)

> But, since in point of empirical fact, I am directly acquainted with the qualia of my own experience, I happen to know (by acquaintance) what the neurophysiologist refers to when he talks about certain configurational aspects of my cerebral processes. (Feigl 1967, 83)

As I understand these passages, one term of such an identity statement would be the kind of term that people typically use to refer to a feature of their experience, and the other term would be a description of certain "configurational aspects" of brain activity that can be discovered using the outward-observation-based methods of neurophysiology. Putting the everyday term first, here are the beginnings of a few such identity statements:

> This itchy sensation in the inner corner of my left eye is...

This auditory sensation produced by the striking of a piano key is...

This orange color sensation produced by that slice of cheese is...

Feigl gives no examples of the neurophysiological terms needed to complete such statements. Presumably he thinks these terms will emerge one day from brain research.

I find it challenging to construct a companion existence statement for such an identity statement because it is not clear to me how Feigl thinks of the common referent of the two terms. I will make three attempts. Here is the first:

> There is a raw feel that consists exclusively of this itchy sensation in the inner corner of my left eye and that also has XYZ configurational aspects that are discoverable by the outward-observation-based methods of neurophysiology.

This is a logical contradiction. If the raw feel consists *exclusively* of the itchy sensation, then it does not also have outwardly observable configurational aspects.

Here is a second attempt, which avoids the contradiction by replacing "consists exclusively of" with "includes":

> There is a raw feel that includes this itchy sensation in the inner corner of my left eye and that also has XYZ configurational aspects that are discoverable by the outward-observation-based methods of neurophysiology.

This statement treats the "raw feel" as a complex thing that has an inwardly observable aspect (the itchy sensation) and also an outwardly observable aspect (the neurophysiological configuration). Such a thing would seem to lack natural definition in the same way as Davidson's "events" and Smart's "brain processes." In any case, like Davidson's identity statements that target complex "events," a two-term identity statement that targets a complex "raw feel" would be compatible with the quadrants view.

Here is my third attempt, which I think comes closest to Feigl's thinking:

> There is a raw feel that consists exclusively of this itchy sensation in the inner corner of my left eye, which one can also refer to using a description of XYZ configurational aspects that are discoverable by the outward-observation-based methods of neurophysiology.

Understood in this way, Feigl's theory involves a magical exception to the normal rules of reference. As I noted in my discussion of Smart's theory, a descriptive expression normally succeeds in referring to something by specifying conditions that the referent satisfies. How then does a description of XYZ configurational aspects that are discoverable by the outward-observation-based methods of neurophysiology manage to refer to a raw feel *that does not have any of these configurational aspects?* Moreover, if there is something that does have XYZ configurational aspects, how does the neurophysiological description manage *not* to refer to *that* rather than to a raw feel? On this interpretation of Feigl's theory, he seems to pin a description of outwardly observable configurational aspects of the nervous system onto an inwardly observable "raw feel" by arbitrary fiat. Smart takes this liberty with the sensation term of his identity statements, declaring that all sensation descriptions refer to "brain processes" that have no inwardly observable properties. Feigl takes it with the neurophysiological term of his identity statements, declaring that some neurophysiological descriptions refer to "raw feels" that have no outwardly observable properties. Both declarations violate the rule that a descriptive referring term refers to what satisfies the description.

Here is another aspect of the same problem. A person in a deep, dreamless sleep has a rather active brain, but no raw feels. Also, the brain of a waking person does many things that are typically not associated with raw feels, such as regulating the secretion of various hormones. Therefore, it is a consequence of Feigl's identity theory that some neurophysiological descriptions of configurational aspects of the brain refer to raw feels and some do not. Feigl acknowledges this point:

> ... it is plausible that *only certain types of cerebral processes in some of their (probably configurational) aspects* are identical with the experienced and acquaintancewise knowable raw feels. (Feigl 1967, 90; my emphasis)

But this raises a thorny question that Feigl does not address. What do all the neurophysiological descriptions of configurational aspects of the brain that do not refer to raw feels refer to? One possible answer is that they refer to configurational aspects of the brain that actually satisfy the descriptions. This creates a bizarre inconsistency between neurophysiological descriptions that refer to what satisfies them and neurophysiological descriptions that refer to raw feels that do not satisfy them. Another possible answer is that neurophysiological descriptions that do not refer to raw feels do not refer to anything. Since there are no raw feels for them to refer to, they simply point into the void. This is simply weird.

In sum, I see no way to understand Feigl's identity theory that makes it a plausible challenge to the quadrants view. On one reading, the theory is self-contradictory. On a second reading, the theory seems to involve entities that lack natural definition, but it is in any case compatible with the quadrants view. On a third reading, the theory violates the normal rules of reference.

Feigl makes an additional claim in connection with his identity theory that conflicts with the quadrants view in an important way. He claims that his identity statements express *"twofold access* or *double knowledge"* with respect to raw feels, where the two types of access are "introspection" (inward observation) and "the science of neurophysiology" (outward observation) (Feigl 1967, 80; his emphasis). According to the quadrants view, there can be double knowledge of a whole living human being, in the sense that some properties of a human being are encountered through outward observation and some other properties are encountered through inward observation, but there is no *property* that can be encountered in both ways. In particular, features of experience are often inwardly observable but never outwardly observable.

As far as I can tell, Feigl says nothing to justify this double-knowledge claim. He simply makes it in order to flesh out his identity theory, which has the problems I have already noted. Against this double-knowledge claim, I cite the plausibility arguments of chapter 7 for the claim that electronuclear fabric has properties of a kind that cannot have evidentiary effects on the spatial transmissions. I also cite the fact that neurophysiologists do not generally have the impression that their outward observations of a living animal's brain give them access to the animal's "raw feels."

In chapter 11 I criticized Feigl's view that the pursuit of outward-observation-based scientific research is somehow inseparable from the belief that nothing real is "un-get-at-able" using outward-observation-based tools and methods (1967, 33–34). His idea that there can be "twofold access" to raw feels is a special case of this more general belief. Both the general belief and this special case of it are expressions of an unjustified and implausible outwardism.

It is interesting to compare the ways in which Feigl and Smart seek to establish the hegemony of outward observation. Feigl accepts the existence of inwardly observable "raw feels" and argues that they are also in some sense outwardly observable, making "double knowledge" of them possible. Smart denies the existence of inwardly observable "raw feels" and maintains that everyday talk about sensations is the reporting of "brain processes" that have no inwardly observable properties. In Feigl's theory, the set of inwardly observable properties is a subset of the set of outwardly observable properties. In Smart's theory, the set of inwardly observable properties is the empty set.

<>

David Papineau says that his identity theory involves "conceptual dualism." On the one hand, there are concepts used in the physical sciences. On the other hand, there are "phenomenal concepts," which are associated with inward observation. The identity statements that he envisions contain one referring term that expresses a physical-science concept and one referring term that expresses a phenomenal concept. The common referent of the two terms, he says, is "a physical property." The following sentence gives an example of how he thinks this might work:

> Suppose we have some theory which identifies pain, say, with some physical property, like the firing of nociceptive-specific neurons in the parietal cortex. (Papineau 2002, 141)

Here, the word "pain" is said to express a phenomenal concept, the phrase "the firing of nociceptive-specific neurons in the parietal cortex" is said to express a physical-science concept, and these two terms are said to have a common referent, which is a certain "physical property."

Someone who speaks of a neuron "firing" is using a simple metaphor for a very complex process. Here is a sketch of what happens when a neuron "fires." A typical neuron has a long, branching appendage called an axon that makes contact with many other neurons. In addition, a typical neuron maintains different concentrations of sodium and potassium ions inside and outside its cell membrane, giving rise to a voltage difference across the membrane. In this respect it resembles a battery. When the cell is stimulated in an appropriate way, pores open in the axon membrane near the cell body, allowing ions to rush through the membrane and the trans-membrane voltage to change abruptly at that location. As the pores close at one location, pores open at the next location along the axon. In this way, a wavelike disturbance travels down the axon that consists of ions moving perpendicular to the axon. It's like a human wave in a sports stadium: the wave travels horizontally through the crowd but it consists of human bodies moving vertically. A disturbance like this can travel down an axon again and again at frequencies up to 100 times per second, which invites comparison to a machine gun that is being fired.

Going behind the "firing" metaphor in this way exposes one problem with Papineau's proposal: there is no naturally bounded entity to serve as the referent of the phrase "the firing of nociceptive-specific neurons in the parietal cortex." How does the referent of this phrase differ from the referent of the phrase "nociceptive-specific neurons in the parietal cortex while they are firing"? Presumably there are many differences. A neuron that is in the process of firing contains cellular nuclei, chromosomes, mitochondria, ribosomes, and many other cell parts that would seem not to be part of "the firing." But then what does "the firing" include? There is no natural boundary inside a firing nerve cell that enables one to answer this question in a nonarbitrary way. Does "the firing" include all the properties of the rushing ions, including their mass? This would give "the firing" a weight, which sounds odd. But if ion mass is not included, which properties of the ions are included and why? What about electric charge, velocity, shape, size, and temperature? What about the complex processes in the axon membrane that open and close the pores, and the various properties of the molecules that surround and control the state of the pores? "The firing" of a neuron is another ill-defined entity, much like Davidson's "events" and Smart's "brain processes." And if there is no well-defined referent for the term "the firing of nociceptive neurons in the parietal cortex," then, a fortiori, there is no well-defined referent for this term and the term "pain" to share.

Incidentally, a similar situation obtains in many cases where you take a verb—let's use the make-believe verb "to voo" as an example—and form the phrase "the vooing." The reason is that we typically apply verbs based on salient characteristics of objects, which happen to be inseparable from indefinitely many other characteristics that are less apparent. For example, there is a salient difference between a person who is walking and a person who is not walking, but the phrase "the walking" has no definite referent. Which properties of which muscles and nerves does "the walking" include? Does it include the walker's visual experience as he walks, or his thoughts about his route and destination, or the way his clothing flaps as he moves his arms and legs? One can pose a comparable set of questions-without-answers for "the sitting," "the talking," "the washing," and many other phrases that have this grammatical form.

A second problem with Papineau's proposal is that it conflicts with a belief that is hard to shake. According to Papineau's identity theory, the common referent of the two terms is both outwardly observable and inwardly observable. Much like Feigl, Papineau imagines that there are properties that one can have double knowledge of or twofold access to—although he does not use these expressions. His conception of the properties that can serve as the common referents seems rather different from Feigl's, but he shares the belief in twofold access. As a consequence, Papineau's identity theory conflicts with the everyday impression that one typically uses the word "pain" to refer to something that is not outwardly observable and that is very different from the velocities and electric charges of sodium and potassium ions or the opening and closing of pores in an axon membrane. For Papineau's identity theory to be correct, this everyday understanding of the word "pain" must involve an illusion of some sort. Papineau needs to give some reason to think that there is an illusion here. He does present an argument that is intended to show this. However, the argument is defective. Because the argument concerns the causation of human behavior, which is the subject of the next chapter, I criticize it there.

<>

Christopher Hill's identity theory is especially vague about the kind of thing that is supposed to serve as the common referent of an identity statement's two referring terms. He speaks of "brain processes" but also of "states." These do not sound like the same kind of thing, and Hill does not explain what he means by either term.

Hill argues for his identity theory on the ground that certain identity statements provide "the best explanation" of certain correlations between inward and outward observations. The following passage illustrates his thinking:

> Suppose, for example, that conscious experiences of a certain
> kind φ turn out to be correlated with brain processes of kind

ψ. Surely, if someone were to ask for an explanation of this correlation, it would be perfectly appropriate to respond by saying, "The correlation obtains because *being a conscious experience of type φ* is the very same thing as *being a brain process of type ψ*." (Compare: "Miss Lane, why does Clark Kent always turn up in the same places as Superman?" "Because, Jimmy, Clark *is* Superman.") (Hill 1991, 24; his emphasis)

Hill is right that some correlations can be explained by two-term identity statements. However, he describes the relation between a correlation and an explanatory identity statement in a way that misrepresents how such an explanation works. The correlation in need of an explanation is between two sets of observed properties, and the common referent of the explanatory identity statement is a third item that possesses all the properties. In the Superman scenario, for example, there is a correlation between 1) observations of a mild-mannered reporter in a gray suit and 2) observations of a flying superhero in a colorful cape, and the common referent of the explanatory identity statement is 3) an individual who dresses in these two different ways as the occasion requires. Hill's sketchy description of this scenario obscures this three-pronged configuration, creating the impression that the gray-suit, the colorful-cape, and the individual who wears them are somehow all the same thing. Incoherently, the existence of a correlation between *two* sets of observed properties is affirmed and denied in the same breath. Clark Kent is Superman, but Clark Kent's gray suit is not Superman's colorful cape. Hill makes the same mistake in explaining correlations between inwardly observed experiential properties and outward observed properties of the brain. If such a correlation could be explained by an identity statement, the common referent of the two terms of the statement would be some third thing that has both an inwardly observable property and an outwardly observable property—like one of Davidson's complex "events." Such an identity statement would not support Hill's "materialist" view that the inwardly observable property is the outwardly observable property.

Hill is also wrong that identity statements provide the best explanation of the correlations that have been discovered between inwardly observable experiential properties and outwardly observable properties of the brain. In fact, given that a true identity statement must have two fully specified referring terms that have a naturally well-defined common referent, one cannot explain these correlations with identity statements at all. The best explanation of these correlations is causal. Specifically, it is the causal explanation that is given by the two-quadrant foundation hypothesis of the quadrants view. Each experiential property is the joint product of outwardly observable properties and remote properties that belong to an indissoluble property package. The causal participation of outwardly observable properties explains the correlations between the experiential property and the participating outwardly observable properties. The causal participation of remote properties explains why the properties that are seen to be correlated cannot be seen to be related to each other in an intelligible way.

<>

Paul Churchland builds an identity theory around what he calls *sensory coding vectors*. A sense organ typically has receptor cells of several types. When the organ is subjected to a given stimulus, receptor cells of each type respond differently. One of the differences is that the receptor cell axons that lead to the brain "fire" at different frequencies, up to a maximum of about 100 times per second. A sensory coding vector is a set of numbers that specifies the respective firing frequencies of the several types of receptor cell in response to a given stimulus. For example, if the human tongue has taste receptors of four types, then the sensory coding vector for a given stimulus to the tongue is a set of four firing frequencies. One possible set of values is (20, 50, 5, 85). Churchland writes:

> What is interesting is that subjectively similar tastes turn out
> to have very similar coding vectors. (Churchland 1988, 148)

For example, apricots taste similar to peaches, and the sensory coding vectors associated with eating apricots and eating peaches are in the same numerical neighborhood (Churchland 1995, 23). The sensory coding vectors associated with sweet tastes form a numerical cluster. The sensory coding vectors associated with bitter tastes form another numerical cluster. And so on. This is indeed interesting—a significant research result. But Churchland goes astray in claiming that such findings support identity statements:

> Here there is definite encouragement for the identity theorist's
> suggestion that any given sensation is simply identical with a

set of spiking frequencies in the appropriate sensory pathway.
(Churchland 1988, 148)

There are three problems with this suggestion.

First, a "set of spiking frequencies" or a "sensory coding vector" seems to be an abstract way of thinking about what is happening in a person's nervous system, not what is actually happening there. The actual happening includes membrane pores opening, ions rushing through the open pores, membrane pores closing, and many other ultra-fast changes in the electronuclear fabric of nerve tissue.

Second, Churchland seems to be in the same muddle as Hill concerning the relation between correlation and identity. The research result that Churchland reports is a correlation between two sets of observed properties—inwardly observed taste sensations and outwardly observed neuron firing frequencies. This very fact shows that the sensations are *not identical with* the firing frequencies. As I explained in my discussion of Hill's theory, there are cases in which a correlation can provide support for an identity statement, but there is no case in which a correlation can provide support for an identity statement that asserts the oneness of what must be two in order for the correlation to exist. There has to be a three-pronged configuration—two sets of properties and a third item that is the bearer of both sets of properties.

The following sentence is interesting for the way it fudges the relation between identity and analogy:

> Whether or not mental states turn out to be physical states of
> the brain *is a matter of* whether or not cognitive neuroscience
> eventually succeeds in discovering systematic neural analogs
> for all of the intrinsic and causal properties of mental states.
> (Churchland 1995, 206; my emphasis)

The phrase "is a matter of" wrongly suggests that the existence of systematic analogies is somehow tantamount to the truth of certain identity statements. It is true that everything that exists is systematically analogous to itself, but it is also true that there are many systematic analogies between different things, due to the patterned ways in which causation works. My right hand is systematically analogous to my left hand. English is systematically analogous to French. A row of trees on the far side of a calm lake is systematically analogous to its reflection in the water. A reproduction of a painting is systematically analogous to the original. One spiral strand of a double-helix DNA molecule is systematically analogous to the other strand. A ball-and-stick model of a protein molecule is systematically analogous to the molecule. The orbit of Mars is systematically analogous to the orbit of the earth. A digital recording of a piece of music is systematically analogous to the live musical performance, to the written musical score, to the musical thoughts of the composer, and perhaps even to some other piece of music that the composer was influenced by. This variety of examples, illustrating different types of causal connection, is intended to drive home how commonplace causally produced systematic analogies between different things are. To say that every systematic analogy indicates an identity is as far from the truth as saying that every red object is a STOP sign. Whatever systematic analogies exist between inwardly observable experience and outwardly observable brain properties could be due to patterns of causation. And this is in fact the case, according to the quadrants view.

The third problem with Churchland's theory is the arbitrariness of focusing exclusively on the firing frequencies of neurons. There are other outwardly observable properties of brains that are just as closely correlated with features of experience as firing frequencies are. The movement of electrical disturbances along axons is triggered by changes in the neuron cell bodies. As the activity reaches the end of an axon, it triggers the release of neurotransmitter molecules into a synapse. Released neurotransmitter molecules diffuse across synapses and trigger chemical processes at the receptor sites of other neurons. There is increased blood flow to neurons that are involved in such activities. If there were a valid inference from correlation or analogy to identity statements, one could infer that a given experiential taste sensation is "simply identical with" outwardly observable properties of all these types, which by the transitivity of identity would all be identical with one another. This is plainly absurd.

It seems to me, incidentally, that many who write about brain and mind focus excessively on neuron firing. This includes advocates of identity theories as well as many who do not advocate identity theories. As far as I can determine, disturbances traveling along axons tend to monopolize people's attention only because they are relatively easy to detect and measure. I have no doubt that this axonal activity plays an important role in the generation of experience, but it seems to me that its role could be more modest than many writers presume it to be. For example, maybe it merely performs a coordination function, akin to email and telephone calls in the work life of a large organization. According to this suggestion, the climactic step in the generation of animal experience takes place in neuron cell bodies, which do whatever they do *in concert* thanks to the axonal traffic travelling incessantly amongst them.

<>

Perhaps you are thinking that the problems of the six identity theories that I have discussed in this chapter can be avoided by constructing a better identity theory. If so, recall the central claim of the previous chapter: there is no such thing as an identity theory in the sense of a theory that *requires* two-term identity statements for its expression. For every true two-term identity statement about the real world, there is a companion existence statement that covers all the ground covered by the identity statement, and more. Therefore, if there is any threat to the quadrants view from an identity theory, the same threat can be cast in the form of companion existence statements. If there are plausible existence statements that conflict with the quadrants view, what are they?

22

The Quadrants and Human Behavior

This chapter sketches an account of the causation of human behavior that uses key features of the quadrants view. It then assesses the plausibility of this account relative to some well-known alternatives. To set the stage, I begin with some thoughts about causation in general.

<>

Many writers maintain that all that happens is predetermined in every detail. All that has ever happened was predetermined before it happened, and all that ever will happen is predetermined now. This includes every little thing that human beings will ever think, feel, say, or do. According to this view, there is an extensive regime of "causal laws" that fixes the world's future course completely when applied to the present state of things. I don't know of any way to show that this view is false. Moreover, I suspect that there is no way to show that this view is false, because the world takes a single course and a determinist can always claim that it had to take this course. On the other hand, I have never come across a good reason to think that this view is true. It's a notion with no support. Some writers worry whether it makes sense to hold people morally responsible for what they do if their behavior is predetermined in every detail. In my view it makes no sense to worry about this in the absence of any good reason to believe that human behavior is so thoroughly predetermined.

My guess is that there is some openness in the way things happen. I think there are some rigidly deterministic "causal laws," but I don't think there are enough of them to predetermine every detail. I also don't think that all causation involves rigidly deterministic "causal laws." Much causation is a matter of constraint and influence that falls short of strict determination, leaving room for undetermined happening at the margin.

One reason that is often given for belief in strict determinism is the predictive power of modern science. This thought involves a leap way beyond what the available evidence justifies. A fair assessment of our current ability to predict the future yields a mixed picture of the predictable and the not-so-predictable. There is no way to tell how much of the unpredictability is due to ignorance and how much is due to a partial openness in what we are trying to predict.

I think another reason why some people find the doctrine of strict determinism attractive is its simplicity. Everything that happens is determined in advance, period. That is a very simple message. On my view, the actual nature of electronuclear happening is more difficult to describe, perhaps impossibly difficult. The past grows longer in a constrained yet somewhat open way. But how exactly do constraint and openness share the field? I have no answer to that question.

Most people have a sense that their own future behavior is somewhat open. I agree with those who are leery of strict determinism for this reason, but I disagree with those who see the partial openness of human behavior as the only exception to strict determinism. Was the entire history of life on earth, including every genetic mutation, every adventure of every animal, and the blooming and fading of every flower, already predetermined before the solar system formed? Is the singing of every bird, the fluttering of every leaf, and the birth and bursting of every bubble, completely predetermined for the rest of time? Maybe so, but I see no reason to believe it. In my view, the partial openness of human behavior is of a piece with the partial openness of electronuclear happening in general.

Nothing that follows depends on the doctrine of strict determinism being false. If you are a strict determinist, you can understand my account of the causation of human behavior in a strictly deterministic way. But the account is also consistent with the looser and more complicated conception of causation that I favor.

A final general point, and a very important one, is that there is nothing in my understanding of causation that limits the types of properties that can play causal roles. In particular, a property might be able to participate in electronuclear causation whether it is experiential or nonexperiential, and whether it is outwardly observable or inwardly observable or neither.

Accordingly, the causation that takes place in a human electronuclear system could involve properties in any or all of the four quadrants.

<>

Taken together, these general points suggest the following sketch of the causal run-up to any given instance of human behavior. The behavior issues from a complex and continuous path of influence, in general not completely deterministic, that involves the brain, motor neurons, and muscles. The part of this path that is in the brain includes outwardly observable properties, experiential properties, and remote properties, all working together. The part of this path that is in the motor neurons and the muscles includes outwardly observable properties and remote properties, but no experiential properties. The distinctive feature of this account is *multi-quadrant causation*, by which I mean the simultaneous and joint participation of properties in two or more quadrants. In chapter 10 I described the way outwardly observable properties, remote properties, and in some cases immediately preceding experiential properties jointly produce experiential properties. That is also multi-quadrant causation. The overarching point is that a great deal of causation in this world is multi-quadrant causation involving outwardly observable properties, remote properties, and experiential properties in various combinations. The multi-quadrant causation of human behavior exemplifies this general pattern.

There is a weak resemblance between my hypothesis of multi-quadrant causation and Ted Honderich's hypothesis of causation by "psychoneural pairs" (Honderich 1988, Honderich 2005). A psychoneural pair, in Honderich's conception, is a causal unit that consists of a "mental event" and a "neural event" that are united by "psychoneural intimacy." The "mental event" is experiential and the "neural event" has "only neural properties, i.e. electrochemical properties" (Honderich 2002, glossary). This sounds like a description of multi-quadrant causation involving experiential properties and outwardly observable properties. There are many differences between Honderich's view and mine. Honderich is a strict determinist. He says nothing of remote properties. He thinks there are naturally bounded "events" and that causation is due to these "events." He uses the word "neural" in an outwardly skewed way. The resemblance I am noting concerns only the idea that causation can involve the simultaneous participation of properties of different types.

This idea of heterogeneous multi-property causation is notably absent from the work of most writers on this subject, including all of the other writers that I discuss in this chapter. Most writers neither advocate nor attack this idea, but simply fail to mention it, as if it never enters their thoughts.

I turn now to the comparison of my hypothesis of multi-quadrant causation of human behavior with some well-known conflicting accounts. I begin with epiphenomenalism, focusing on the arguments for it made by T. H. Huxley and Jaegwon Kim. Then I consider the accounts of David Papineau, John Searle, and Paul Churchland, which are neither multi-quadrant nor epiphenomenal.

<>

In chapter 2 I distinguished two variants of epiphenomenalism—a body-plus variant and a body-only variant. According to body-plus epiphenomenalism, all of a person's experiential properties belong to a "mental steam" emitted by the brain that plays no role in the causation of the person's behavior. According to body-only epiphenomenalism, all of a person's experiential properties are properties of the person's body that play no role in the causation of the person's behavior. The shared claim that makes these views variants of epiphenomenalism is that experiential properties play no role in the causation of behavior.

I think there is a strong case against epiphenomenalism, consisting of three separate arguments. One argument is aimed specifically at body-plus epiphenomenalism, one argument applies to both variants, and one argument is aimed specifically at body-only epiphenomenalism. There are thus two arguments against each variant.

The argument that specifically targets body-plus epiphenomenalism is that it is difficult to make sense of the "mental steam" body-plus view, with or without epiphenomenalism. The details are in chapter 2.

The argument against epiphenomenalism of any type can be found in most discussions of the subject. It is that people have the overwhelming impression that many of their experiential properties play crucial roles in motivating and shaping their behavior. I find statements such as the following hard to doubt. My visual experience plays a crucial role in guiding my movements while driving my car or playing tennis. My auditory experience plays a crucial role in shaping my responses when people speak to me. The thinking experience involved in making a decision plays a crucial role in leading me to carry out the decision. Feelings of hunger, fear, anger, and curiosity play a crucial role in motivating me to act in appropriate ways. If none of this is true, then I am the victim of a bizarre illusion.

311

The argument that specifically targets body-only epiphenomenalism is that the exclusion of experiential properties from participation in causation seems arbitrary. In the case of body-plus epiphenomenalism, Huxley's steam engine metaphor gives one a way to understand how experiential properties could be causally idle: they are properties of a "mental steam" that somehow gets separated from the busy brain. But body-only epiphenomenalism lacks this explanatory resource. If experiential properties are as integral to the brain as any other properties are, then it is difficult to understand how they could be causally quarantined, so to speak. There are no moats or firewalls separating the brain's experiential properties from its other properties, so what is there to prevent the experiential properties from participating in the causation? The exclusion of experiential properties from participation in causation makes some sense if they belong to a "mental steam" that has been puffed away from the electronuclear system, but it makes no sense if they are integral properties of that system.

Huxley's argument for epiphenomenalism is extremely weak. He presents several examples in which a living organism does something while in a presumptively unconscious state (Huxley 1896b). The examples include experiments with frogs that have had much of their brains removed. These specimens are completely immobile as long as they are left alone, but they swim if they are placed in water and they walk or jump if they are prodded in the rump. Huxley also describes the case of a wounded French soldier who has periodic spells during which he is seemingly unaware of being spoken to or being pricked with pins, but is nonetheless able to go through some of the habitual routines of his daily life. He also mentions "somnambulism" (sleepwalking) and "mesmerism" (hypnosis). Concerning the wounded French soldier, whom he calls F—, Huxley notes that "it is impossible to prove that F— is absolutely unconscious in his abnormal state" (1896b, 235). I think the same caveat applies to the surgically simplified frogs, but let us assume for the sake of argument that these are all genuine cases of organisms moving in various ways while completely devoid of experience.

On the strength of these examples, Huxley surmises "that the machinery which is competent to do so much without the intervention of consciousness, might well do all" (1896b, 226). This comment blurs the distinction between a speculation and a well-supported conclusion. Huxley's set of examples is not a representative sample of the full range of human and animal behavior. It includes only simple cases of reflex and habit. A plausible alternative to Huxley's hypothesis is that significant differences in behavior are accompanied by significant differences in how the behavior is caused. According to this alternative hypothesis, there might well be some relatively simple cases of reflex and habitual behavior, such as those Huxley describes, whose causation does not involve any experiential properties, but the more complex behavior that seems to have experiential properties participating in its causation does indeed have experiential properties participating in its causation. There is nothing in Huxley's extremely limited set of examples that favors epiphenomenalism over this alternative.

Kim's argument for epiphenomenalism calls for a longer discussion. I start by agreeing with Kim's claim that human behavior is not normally "causally overdetermined," in the sense of being doubly caused along two mutually independent paths. Kim explains causal overdetermination as follows:

> In standard cases of overdetermination, like two bullets hitting the victim's heart at the same time, the short circuit and the overturned lantern causing a house fire, and so on, each overdetermining cause plays a distinct and distinctive causal role. The usual notion of overdetermination involves two or more separate and independent causal chains intersecting at a common effect. (Kim 2005, 48)

Papineau, who uses this same claim in a different argument, gives the example of "the death of a man who is simultaneously shot and struck by lightning" (Papineau 2002, 18). I think it is safe to say that human behavior is rarely if ever caused in such a manner. The final causal run-up to any instance of human behavior occurs in a unified manner in the agent's brain, motor neurons, and muscles. It's true that if you go back in time more than a fraction of a second, there may be a variety of converging influences coming from the agent's eyes, ears, memory, and so on. However, such influences are not whole causes that produce the behavior independently of one another; they are causal tributaries that come together in the brain before the final causal run-up. With this basic point agreed to, I will set aside the bizarre possibility of widespread double causation of human behavior and focus on the nature of the unified causation that takes place in the brain, nervous system, and muscles. This is where views differ.

Kim makes extensive use of the phrase "mental causation," by which he seems to mean *purely* mental causation—causation by mental properties alone. He makes an argument concerning "mental-to-physical causation" and a different argument concerning "mental-to-mental causation." I will focus on the mental-to-physical case.

The core of Kim's argument concerning "mental-to-physical causation" is a general principle that he calls "the causal closure of the physical domain" or, more briefly, "physical causal closure." Here are some of his statements of this principle:

> If a physical event has a cause at t, then it has a physical cause at t. (Kim 2001, 276)

> If a physical event has a cause that occurs at t, it has a physical cause that occurs at t. (Kim 2005, 43)

> If a physical event has a cause (occurring) at time t, it has a sufficient physical cause at t. (Kim 2006, 195)

"Physical cause" is another ambiguous "physical" phrase, which I omitted from the discussion in chapter 9 in favor of discussing it here. It could be used to mean a cause that has at least one physical-science property. However, what Kim seems to mean by it is a cause that consists entirely of physical-science properties. I gather this from the statement "According to this principle, physics is causally and explanatorily self-sufficient" (Kim 2005, 16), as well as from the way he uses the principle in his argument. Applied to human behavior, this principle has the corollary that any motion of a person's arms, legs, vocal cords, or other body parts that has a cause at time t has a cause at time t that consists exclusively of physical-science properties.

Kim's argument combines his principle of physical causal closure with the claim that human behavior is not causally overdetermined. The logic of the argument is simple. If an instance of behavior that has a cause at time t has a physical cause at time t (the principle of physical causal closure), and if the behavior does not have both a physical cause and a mental cause at time t (no causal overdetermination), then the behavior cannot have a mental cause at time t. Therefore, any mental properties that exist at time t are not involved in the causation of the behavior. Since this holds for all times t, the conclusion is complete epiphenomenalism.

In making this argument, Kim does not mention multi-quadrant causation or Honderich's theory of causation by psychoneural pairs as other-than-epiphenomenal possibilities that need to be ruled out. He seems to be under the impression that epiphenomenalism and purely mental causation constitute a logically exhaustive set of alternatives. Although this is a gap in his argument, it appears that he is in a position to rule out multi-quadrant causation and causation by psychoneural pairs using an argument that is even simpler than the one he uses against purely mental causation. To rule out multi-quadrant causation and causation by psychoneural pairs, he does not need to say anything about causal overdetermination; all he needs to say is that these hypotheses are incompatible with the principle of physical causal closure. If an instance of behavior that has a cause at time t has a cause at time t that consists exclusively of physical-science properties, then there is no room for experiential and/or remote properties to participate simultaneously in the causation. Any properties that are not physical-science properties must be causally idle.

Kim's principle of physical causal closure is doing a huge amount of work here. It rules out Honderich's idea of causation by psychoneural pairs and my hypothesis of multi-quadrant causation all by itself, without the need for any other premises. And so we come to a key question: is there any reason to believe this principle? All the points that Kim makes to justify it can be found in the following three passages:

> Most philosophers, including anyone who considers himself or herself a physicalist of any kind, accept physical causal closure. If the closure should not hold, there would be physical events for whose explanation we would have to look to nonphysical causal agents, like spirits or divine forces outside spacetime. That is exactly the situation depicted in Descartes' interactionist dualism. If closure should fail, theoretical physics would be in principle incompletable, and it is fair to say that research programs in physics, and the rest of the physical sciences, presuppose something like the closure principle. (Kim 2006, 195)

> ... to give up this principle is to acknowledge that there can in principle be no complete physical theory of physical phenomena, that theoretical physics, insofar as it aspires to be a complete theory, must cease to be pure physics and invoke irreducibly non-physical causal powers—vital principles, entelechies, psychic energies, elan vital, or whatnot. (Kim 1993, 356)

If you reject this principle, you are ipso facto rejecting the in-principle completability of physics—that is, the possibility of a complete and comprehensive physical theory of all physical phenomena. For you would be saying that any complete explanatory theory of the physical domain must invoke nonphysical causal agents. (Kim 1998, 40)

As explained in the discussion that follows, there is no genuine justification here.

All three of these passages claim that if the principle of physical causal closure was not true, people would not be able to "complete" theoretical physics and/or physics in general. This may be the case, but it provides no support for the principle. Believing as I do that theoretical physicists are clawing at a world that is partly beyond comprehension, the suggestion that a "complete physical theory of physical phenomena" or a "complete and comprehensive physical theory of all physical phenomena" is impossible sounds exactly right to me. One cannot justify a principle by showing that its negation has a plausible implication.

The first two passages mention certain imaginable exceptions to the principle of physical causal closure that involve entities that we have no reason to believe in. The first passage mentions "spirits or divine forces outside spacetime." The second passage mentions "vital principles, entelechies, psychic energies, elan vital, or whatnot." If there is an argument here, it must be that the only alternative to accepting the principle is to believe in such outlandish entities. But this is clearly not the case. Even Kim does not believe this, because according to his argument for epiphenomenalism one alternative to accepting the principle is to believe in purely mental causation by familiar experiential properties. Other imaginable alternatives involve Honderich's idea of causation by psychoneural pairs and my hypothesis of multi-quadrant causation. To justify a principle by eliminating imaginable alternatives, one has to eliminate all the imaginable alternatives, not just the outlandish ones.

The first passage makes a couple of additional points. One is that "most philosophers" believe this principle. This may be so, but it provides no justification. The other is that "research programs in physics, and the rest of the physical sciences, presuppose something like the closure principle." This sounds like a claim that the very possibility of scientific research depends on the truth of this principle. This is emphatically not so. The nature of causation is a subject for scientific investigation, not something that one must assume in advance! Research in the physical sciences can go forward if there is purely mental causation, causation by psychoneural pairs, multi-quadrant causation, or even occasional causal incursions from "spirits or divine forces outside spacetime." The results of some research might depend on which if any of these is true, but the ability of scientists to conduct the research does not.

In sum, Kim's argument for epiphenomenalism rests on a principle that he fails to justify. It also seems to rest on the false notion that the only possible alternative to epiphenomenalism is purely mental causation. Another possible alternative to epiphenomenalism is multi-quadrant causation that involves experiential properties. For these reasons, Kim's argument for epiphenomenalism is unsound.

Body-only epiphenomenalism is logically consistent with the quadrants view. One could hold that human beings and other conscious animals are electronuclear systems that have outwardly observable properties, remote properties, and experiential properties, and add to this the claim that the experiential properties (and also the remote properties, if you like) play no role in the causation of behavior. However, given the absence of any good arguments for epiphenomenalism and the strong arguments against it, I consider body-only epiphenomenalism much less plausible than my hypothesis of multi-quadrant causation.

<>

Kim and I do have an important point of agreement: there is no such thing as purely experiential causation. But we arrive at this conclusion in different ways. As I have just explained, I reject Kim's argument, which is based on the unsupported principle of physical causal closure. Here is my argument for the same conclusion. All causation of human behavior takes place in an electronuclear system that consists of organs, cells, and molecules. These things have numerous outwardly observable, nonexperiential properties. It is therefore difficult to imagine how there could be a case of causation in such a system that does not include, at every instant, the participation of some of these outwardly observable, nonexperiential properties. If outwardly observable, nonexperiential properties are *always* involved, purely experiential causation *never* takes place. Whenever experiential properties participate in causation, they do so as part of a multi-quadrant stream that has an uninterrupted outward profile.

Unlike Kim's use of the principle of physical causal closure, this argument does not count against multi-quadrant causation. It distinguishes sharply between purely experiential causation and partially experiential multi-quadrant causation, whereas Kim's principle of physical causal closure counts equally against both of these. Based on this argument, I consider the hypothesis of purely experiential causation of behavior as implausible as epiphenomenalism. These are opposite extremes, which should both be rejected. Multi-quadrant causation that includes participating experiential properties is much more plausible than either of them.

Kim introduces his discussions of "mental causation" with the following question:

> How is it possible for the mind to exercise its causal powers in a world that is fundamentally physical? (Kim 1998, 30)

> How can the mind exert its causal powers in a world that is fundamentally physical? (Kim 2005, 7)

The proper response to this question is to question its presupposition that the world is fundamentally physical. What does Kim mean when he says that the world is fundamentally physical, and is the world fundamentally physical in that sense? If he means that all fundamental properties are physical-science properties, or that all causally active properties are physical-science properties, then there is no reason to think that the world is fundamentally physical. The world is fundamentally itself. Its relation to the outward-observation-based physical sciences is not as close as Kim believes. According to the quadrants view, there is a vast expanse containing electronuclear objects that have properties of various types, including outwardly observable physical-science properties, remote properties, and (in some cases) experiential properties. In this multi-quadrant world, it is "possible for the mind to exercise its causal powers" in the sense that human beings and other animals are sites of multi-quadrant causation that includes participating experiential properties (and perhaps participating evolved remote properties that might also count as "mental").

<>

David Papineau defends his identity theory, which I described in the previous chapter, using an argument that concerns the causation of human behavior. If his identity theory is flawed in the ways that I have said it is, then his argument for it must be flawed too. I explain here why that is indeed the case.

An important component of Papineau's argument is a general principle of causation that is very similar if not identical to Kim's principle of physical causal closure. Papineau states his general principle of causation as follows:

> All physical effects are fully caused by purely *physical* prior histories. (Papineau 2002, 17; his emphasis)

He says that he understands "physical" as "standing roughly for the kinds of first-order properties studied by the physical sciences" (Papineau 2002, 15). He also says that physical properties in his sense are "non-mentally identifiable" and indeed "identifiable non-mentally-*and*-non-biologically" (Papineau 2002, 41; his emphasis). They include properties that are "specifiable in terms of mass, or charge, or chemical structure" (Papineau 2002, 42). All of this suggests that Papineau's general principle of causation is Kim's general principle of causation, or very nearly so.

In the course of explaining this principle, Papineau introduces the word "inanimate" in a dubious way. He writes:

> In fact, I shall understand 'physical' in a somewhat tighter sense in what follows, as 'identifiable non-mentally-*and*-non-biologically', or 'inanimate' for short, rather than simply as 'non-mentally identifiable'. (Papineau 2002, 41; his emphasis)

He goes on to use the word "inanimate" repeatedly as equivalent to "identifiable non-mentally-and-non-biologically." According to this terminology, all inanimate properties are identifiable—a condition that is not part of the ordinary meaning of the word. Yet Papineau does not indicate that he is stipulating a special meaning for "inanimate"; he writes as if he thinks that the word "inanimate" in its ordinary sense is equivalent to "identifiable non-mentally-and-non-biologically." This suggests a tacit assumption on his part that all properties that are inanimate in the ordinary sense are identifiable, and perhaps also a broader assumption that all properties are identifiable. If he is making either of these assumptions, the following two points are in order. First, both of these assumptions are in conflict with the quadrants view, according to which inanimate electronuclear objects have remote properties that are inanimate but not identifiable. Second, Papineau does not give, and does not try to give, any reason to think that either of these assumptions is true.

The argument in which Papineau uses his general principle of causation has three premises, which he numbers and states as follows:

1. Conscious mental occurrences have physical effects.
2. All physical effects are fully caused by purely *physical* prior histories.
3. The physical effects of conscious causes aren't always overdetermined by distinct causes.

(Papineau 2002, 17–18; his emphasis)

Note that the conclusion of Kim's argument is precisely the negation of Papineau's first premise. Their arguments are quite different, despite their use of the same general principle of causation. Here is Papineau's explanation of why his identity theory follows from the conjunction of these three premises:

> Premisses 1 and 2 tell us that certain effects have a conscious cause and a physical cause. Premiss 3 tells us that they don't have two distinct causes. The only possibility left is that the conscious occurrences mentioned in (1) must be identical with some part of the physical causes mentioned in (2). This respects both (1) and (2), yet avoids the implication of overdetermination, since (1) and (2) no longer imply *distinct* causes. (Papineau 2002, 18)

I think Papineau is right that the conjunction of his three premises has this implication. However, I also think that all three of his premises are false. Premise 3 contains the correct idea that human behavior is not typically causally overdetermined, but the premise as a whole is false because it presupposes that premise 1 is true. Premises 1 and 2 both conflict with my hypothesis of multi-quadrant causation, and Papineau's attempts to justify them are seriously flawed, as I will now explain.

Suppose that the causation of human behavior is typically multi-quadrant causation. Then premise 1 wrongly casts the inwardly observable experiential aspect of the multi-quadrant causation as the whole cause at the time it exists, while premise 2 wrongly casts the outwardly observable, physical-science aspect of the multi-quadrant causation as the whole cause. Papineau's argument therefore has much in common with the following manifestly silly argument:

> Serena Williams won today's big doubles match.
> Venus Williams won today's big doubles match.
> Therefore, Serena Williams is Venus Williams.

This would be a compelling argument if Serena had won the match *singlehandedly* and Venus had won the match *singlehandedly*. However, neither of these is the case; this is not how a tennis doubles match works. Likewise, if conscious sensations and intentions cause a certain instance of behavior singlehandedly and outwardly observable physical-science properties cause the same instance of behavior singlehandedly, Papineau would have a strong argument for his identity theory, but there is good reason to doubt that the causation of human behavior works this way. The causation might well involve the concurrent participation of outwardly observable physical-science properties, experiential properties, and perhaps also remote properties. In this case, Papineau's argument is an abstract logical exercise that misses the main point.

Now let's consider what Papineau says to justify premises 1 and 2.

To justify premise 1, he merely appeals to the implausibility of epiphenomenalism, using the example "I walk to the fridge to get a beer, because I consciously feel thirsty" (Papineau 2002, 17). I agree with Papineau that it is very hard to accept the epiphenomenal view that in such a situation a person's feeling of thirst plays no role at all in causing him to walk to the kitchen. But this leaves open the possibility of multi-quadrant causation, in which the feeling of thirst plays a causal role along with other concurrent properties of the agent. When I observe myself getting a drink in response to a feeling of thirst, I have no way to tell whether the inwardly observable feeling of thirst is causing my behavior singlehandedly or as part of a multi-quadrant causal stream. Like Kim, Papineau never considers the possibility of multi-quadrant causation and so makes no attempt to rule it out. For this reason, he does not succeed in justifying the claim of premise 1 that at the time it occurs a "conscious mental occurrence" is the entire cause of the ensuing behavior.

Papineau's attempt to justify premise 2, his general principle of causation, consists of a long discussion of a related yet crucially different question. Instead of discussing *causation*, he summarizes the history of scientific thinking about *forces* and comes to the conclusion that there are no special "mental or vital forces" that operate only in living things (Papineau 2001; Papineau 2002, 232–256). The forces that operate in living things are the same forces that operate outside living things—gravity, electric and magnetic forces, and nuclear forces. Papineau seems to assume that this conclusion about forces is equivalent to his general principle of causation. But it is not. I know of no reason to think that there are special "mental or vital forces." Let's assume for the sake of discussion that there are none. Papineau's general principle of causation could still be false, for two reasons.

First, the effect of applying a net force to an object is a change in the speed and/or direction of the object's motion, but even physical scientists study changes of other types than this. A few examples are the absorption of a photon by a molecule, the emission of a photon by a molecule, the quantum jump of an electron in a molecule to a different energy level, and the mutual annihilation of two colliding particles such as an electron and a positron. Such changes are not caused by the mere application of a force to an object. It seems, then, that causation by the application of a force to an object is only a part of the subject of causation. By limiting his discussion to forces, Papineau overlooks much of the subject that his principle of causation must cover. It is interesting in this connection that Papineau calls his principle "the completeness of physics." This name contains neither the word "cause" nor the word "force" and thus helps to obscure the issue of how causation and forces are related. I see some suggestions in Papineau's work that he holds the view, which dates back to ancient atomism, that all change is motion. If he believes this, it would explain why he thinks he can justify a general principle of causation by discussing only forces. But if this belief is functioning as an additional, unstated premise in his argument, he needs to explicitly justify it, because there are so many apparent counterexamples to it.

Second, force is a complex, incompletely understood, and somewhat mysterious phenomenon. Physicists say that the force between two objects is "carried by" particles that the objects "exchange." For example, the electric force between two protons, two electrons, or an electron and a proton is carried by photons (or virtual photons—I don't understand what the difference is). There seems to be room for remote properties to play a role in these proceedings. If remote properties are involved, even the causation of a change in an object's motion by an applied force would violate Papineau's general principle of causation. Papineau seems to assume that the forces that physicists study involve only identifiable, physical-science properties, but I know of no reason to believe that. Physicists who write about forces may well be describing only the outward profiles of multi-quadrant processes.

Perhaps the familiar expression "the fundamental forces" is a culprit here. The fundamental forces are so named because they are believed to be fundamental among forces. But whether forces are fundamental among all realities is another question. There might be objects and properties of objects that are more fundamental than any force.

Papineau's justification of his general principle of causation appears in a highly compressed form in the following passage by Hilary Putnam:

> In the seventeenth century, people became aware that the physical world is *strikingly causally closed*. The way in which it is causally closed is best expressed in terms of Newtonian physics: no body moves except as the result of the action of some *force*. Forces can be completely described by numbers: three numbers suffice to determine the direction, and one number suffices to describe the magnitude of any force. (Putnam 1981, 75; his emphasis)

This passage has many flaws. It uses the problematic phrase "the physical world" without explanation. Its thumbnail sketch of Newtonian physics sounds more like a sketch of Aristotelian physics. According to Newtonian physics, it is only *changes in motion* that require forces; ongoing motion in a fixed direction at a constant speed requires no force in the Newtonian scheme. The second sentence slides without justification from "cause" to "force," just as Papineau does. The third and final sentence makes force seem much simpler than it actually is, also like Papineau. Putnam claims that forces "can be completely described" by four numbers. But the four numbers that Putnam alludes to do not tell you which force they are describing or anything at all about the objects that project, respond to, or carry that force. Indeed, the three direction numbers do not even describe a direction; all they do is relate the direction of the force to three other directions, which define an arbitrarily chosen system of coordinates. A direction is a *direction*; it is not a set of numbers. Likewise, the magnitude number merely relates the magnitude of the force to another force magnitude that has been arbitrarily chosen as a unit of measurement. It says nothing about the nonmathematical reality of force magnitude. These numbers are useful for computational purposes, but they tell us nothing about the *nature* of force, which is complex, mysterious, and possibly less "physical" than Papineau and Putnam assume.

<>

It is noteworthy that Kim and Papineau attempt to justify what seems to be the same general principle of causation in very different ways. Papineau's justification is all about forces, whereas Kim presents a miscellany of considerations without ever using the word "force." This fact alone is a red flag. If a principle is sound, one might expect those who believe it to believe it for the same reason. If Kim is right that "most philosophers" believe this principle, there might well be other approaches to justifying it among them. But is there an approach that works? I doubt it.

Why is this principle so widely accepted if the justifications given for it are so shaky? The reason, I believe, is that it reflects widespread habitual outwardism. If you think of the world as a realm of outwardly observable physical-science properties, then you are going to think of causation as something that involves only such properties. Papineau uses the telltale phrase "the third-personal, causal world" (Papineau 2002, 48), which suggests that there is an essential connection between outward observation and causation. That is the fundamental mistake. In the same vein, Kim says that "causality is fundamentally a physical phenomenon" (Kim 2005, 55). What does this mean? If it means that causality involves only outwardly observable physical-science properties, then there is no reason to believe it. Causation might well be a plenary reality that involves properties of many types, without regard to how or whether human beings can observe them. According to the hypothesis of multi-quadrant causation, outward observation and the physical sciences give us access only to causation's outward profile.

<>

My main discussion of John Searle is in chapter 13, but for organizational reasons I have saved his account of the causation of human behavior for this chapter. What Searle says on this subject seems self-contradictory to me. On this point I agree with Kim, who concludes a critique of Searle's account with this admonition:

> Searle needs to come up with a reasonable ontology and language of causation to make his central claims about

> the mind-body relation intelligible and consistent. (Kim
> 1998, 49)

The required revision is not minor. There is a fundamental problem with Searle's account that calls for a major change.

In his discussions of this subject, Searle focuses on a person's conscious intention to do something. His favorite example is the relation between a person's conscious intention to raise his arm and the arm going up. This is an extremely narrow focus. Experiential properties of many types, not just intentions, seem to be involved in the causation of behavior. A few examples are feeling thirsty, hearing a spoken request to do something, and seeing an object in front of you that you do not want to bump into. Although the narrowness of Searle's focus is noteworthy, it is ultimately inconsequential because his account has the same problem no matter what type of experiential property it is applied to.

Concerning the arm-raising example, what Searle seems to say is that the upward motion of the arm is wholly caused by the agent's conscious intention to raise the arm, and wholly caused by properties that a neurobiologist can discover through outward observation, but that this does not amount to causal overdetermination because in saying these two things one is merely describing the same causation in two different ways. This part of Searle's account sounds a lot like Papineau's causal argument for his identity theory. However, as we saw in chapter 13, Searle also holds that experiential properties are distinct from the properties that a neurobiologist can discover through the outward observation of a person's brain. He even carries this claim to the obscure extreme of saying that properties of these two types have different "modes of existence" or different "ontologies." Experience has a "subjective" mode of existence, while the properties studied by neurobiologists have an "objective" mode of existence. Putting all this together yields the following incomprehensible result. Causation of an instance of behavior entirely by a conscious intention is causation of that instance of behavior entirely by certain outwardly observable neurobiological properties of the brain, even though the conscious intention is distinct from all these outwardly observable neurobiological properties, to the point of having a different "mode of existence."

I cannot imagine how all of this can be true. It seems to me that an instance of causation by a certain set of properties includes those causally involved properties. Therefore, if there is causation entirely by a set of properties that have a "subjective" mode of existence and causation entirely by a different set of properties that have an "objective" mode of existence, these must be two distinct instances of causation. Conversely, if there is only one instance of causation that involves all of these properties, then it must be causation by a *combination* of "subjective" experiential properties and "objective" physical-science properties, neither of which constitutes the whole cause. In my terminology, it must be a kind of multi-quadrant causation. I cannot make sense of the claim that causation by just a conscious intention is causation by just a distinct set of outwardly observable neurobiological properties. That to me is self-contradictory.

Why does Searle have such confidence in this seemingly self-contradictory view? Part of the answer, I believe, is his reliance on the slippery phrase "level of description." He distinguishes "system-level" descriptions, which describe a system without referring to any of its parts, from "lower-level" descriptions, which refer to parts of a system. He then makes certain blanket claims about the relationship between "system-level" descriptions and "lower-level" descriptions of the same system. The result is fallacious reasoning because the category of "system-level" descriptions includes descriptions of different types, which are related to "lower-level" descriptions in different ways. By treating "system-level" descriptions as a monolithic category, Searle misses these differences.

Here is an example that illustrates this pitfall. Following are three "system-level" descriptions of a certain system:

> My next-door neighbor
> My next-door neighbor's body
> My next-door neighbor's cheerful personality

Following are two "lower-level" descriptions of the same system:

> My next-door neighbor's toes
> My next-door neighbor's red blood cells

How are the system-level descriptions related to the lower-level descriptions? The important point is that this question has no single answer. The phrase "my next-door neighbor" subsumes the two lower-level descriptions. So does the phrase "my next-door neighbor's body." However, the phrase "my next-door neighbor's cheerful personality" does not subsume either of these lower-level descriptions. The reason is that it refers not to the whole system but to a particular property of the system that is only distantly related to the system's toes and red blood cells.

Searle's argument in the following passage seems to depend on ignoring precisely this sort of distinction between different types of "system-level" description:

> When I say that my conscious decision to raise my arm caused my arm to go up, I am not saying that some cause occurred in addition to the behavior of the neurons when they fire and produce all sorts of other neurobiological consequences, rather I am simply describing the whole neurobiological system at the level of the entire system and not at the level of particular microelements. The situation is exactly analogous to the explosion in the cylinder of a car engine. I can say either the explosion in the cylinder caused the piston to move, or I can say the oxidation of hydrocarbon molecules released heat energy that exerted pressure on the molecular structure of the alloys. These are not two independent descriptions of two independent sets of causes, but rather they are descriptions at two different levels of one complete system. (Searle 2004, 208–209)

The description "the explosion in the cylinder" does seem to subsume details about the oxidation of hydrocarbon molecules, just as Searle says. However, the description "my conscious decision to raise my arm" does not seem to subsume any details about outwardly observable properties of the brain, for the same reason that "my neighbor's cheerful personality" does not subsume any details about my neighbor's toes or red blood cells. These are "system-level" descriptions that refer to a particular property of a system, not to a whole system. For this reason, Searle's claim that the two situations are "exactly analogous" is not correct. On the contrary, the two situations are crucially different. If they seem "exactly analogous" to Searle, this is because he is considering them through the sprawling category of "system-level description."

The same point holds for a "system-level" description that refers to any particular feature of experience, such as "my feeling of thirst," "my auditory experience of a request to pass the bread" or "my visual representation of a tree looming in front of me." These are "system-level" descriptions because they do not refer to parts of the system, but they do not subsume any parts of the system because they refer to particular "system-level" properties, not to the whole system. I think Searle is right to insist that there is a real difference between features of experience and outwardly observable properties of the brain (although not a difference in their "modes of existence"). But having affirmed this real difference, he is not free to treat a verbal reference to a feature of experience and a verbal reference to a set of outwardly observable brain properties as alternate descriptions of the same thing. When Searle is not discussing the causation of behavior he stresses the distinction between two "modes of existence," but when he is discussing the causation of behavior the "modes of existence" are off stage and it is all about "levels of description." It feels like a magic trick.

It may be that Searle finds his argument concerning "levels of description" attractive in part because he holds a certain general principle of causation. If so, Searle's general principle of causation is not that causation is fundamentally "physical," as Kim and Papineau would have it, but rather that causation is fundamentally *small-scale*. The way Searle uses the words "all" and "entirely" in the following passage suggests such a principle:

> Of course, the conscious state causes the movement of the arm in virtue of being conscious, because unless it were conscious, it could not cause that movement in that way. But like all higher-level phenomena, such as the explosion inside the car cylinder moving the piston, or the hammer driving the nail, it is causally explicable *entirely* by the behavior of lower level phenomena. (Franken 2010, 216; Searle's emphasis)

If there is a general principle of causation implicit in this passage, Searle never attempts to justify it or even to state it explicitly. As I explained in chapter 2, I think the claim that there is a hierarchy of levels in us is wrong, unless it is intended only in a loose and metaphorical sense. How many "levels" are there between an electron and a man, what are these levels, and how is each level distinguished or separated from the next? Does Searle's apparent principle apply at all size scales down to the lowest level, or only at size scales above a certain level? I don't think there is any good way to answer these questions, or any reason to believe a general principle of causation that presupposes a hierarchy of levels. It seems entirely possible to me that the participation of experiential properties in the causation of human behavior is not explicable at any lower "level" or indeed in any other way. It's just a fact that the multi-quadrant causation of much human behavior includes the participation of various experiential properties.

<>

In his book *Matter and Consciousness*, Paul Churchland presents "four reasons, all directed at the conclusion that *the correct account of human-behavior-and-its-causes* must reside in the physical neurosciences" (1988, 27; my emphasis). Accounts of the causation of human behavior that refer to an agent's thoughts, feelings, sensations, motives, or anything else that one can access inwardly Churchland dismisses as old-fashioned and erroneous "folk psychology."

The four reasons that Churchland gives concern embryology, evolution, and the history of neuroscience. I think they add up to a strong case that the final causal run-up to human behavior occurs in the nervous system—a point that Churchland and I agree on. But I don't think they support his conclusion. According to the quadrants view, the physical neurosciences can provide only a partial account of what goes on in the nervous system because of their dependence on outward observation and the spatial transmissions. Likewise, so-called folk psychology can provide only a partial account of what goes on in the nervous system. Even the combination of the physical neurosciences and so-called folk psychology can provide only a partial account because neither of them deals with remote properties. Churchland says nothing to discredit my hypothesis of multi-quadrant causation or to justify the crucial step from the nervous system to "the physical neurosciences." He doesn't even acknowledge that this crucial step is a step. The reason, I believe, is that he is in the grip of outwardism and therefore doesn't think of this as a step. He harbors the Flatlandish presupposition that the physical neurosciences can provide access to all the properties of the nervous system. But he gives no reason why one should think of the physical neurosciences as providing such comprehensive insight, and I don't believe he can.

<>

I conclude that the hypothesis of multi-quadrant causation of human behavior is more plausible than any of the other views discussed in this chapter. Many of the arguments made by advocates of these other views suffer from their dependence on a general principle of causation that there is no good reason to accept. All the arguments made by advocates of these other views suffer from their failure to consider the possibility of multi-quadrant causation. My challenge to advocates of these other views is to address the hypothesis of multi-quadrant causation of human behavior explicitly. Explain where this hypothesis goes wrong, or else agree with me that it is a plausible alternative to the various accounts that conflict with it.

23

Living with the Quadrants View

There are a couple of reasons why you might find the quadrants view disquieting. This would not be an argument against it, of course. Sometimes the truth is disquieting. But it seems to me that the potential of the quadrants view to arouse unpleasant feelings is actually quite limited.

One crunch point is the incompatibility of the quadrants view with belief in personal immortality. A human electronuclear system is in certain respects like a hurricane—impressive in its prime but limited in time to an arc of birth, growth, peak performance, and disintegration. There is no hurricane heaven. Granted, the thought that death is final can be disturbing. But what are we to do? The quadrants view is by no means the only view that leaves no room for personal immortality. Most of the views discussed in this book, starting with the ancient atomism poetically proclaimed by Lucretius, have this consequence. It is very difficult to fill in the details of a credible world view that permits individual human beings to continue living in an altered form after their blood stops circulating. So perhaps the disquieting thought that death is final is something that we should humbly accept. It should be some consolation that this is a thought that most people do not have very often or for very long. On a typical day, I spend more time brushing my teeth than thinking about my eventual death. This is because it's a simple, bare-bones, almost boring thought, which is usually quickly displaced by one of the many more interesting matters that compete for a person's attention.

The other potential source of malaise is that the quadrants view eliminates all hope of understanding how experience and consciousness come into existence. If the generation of experience depends crucially on remote properties that no feasible electronuclear being can have cognitive access to, then the project of explaining how experience is generated is a nonstarter. Scientists will be able to specify in ever greater detail numerous correlations between particular experiential properties and the outwardly observable properties that play a role in producing them. Some of these discoveries will be fascinating and many of them will lead to significant advances in medicine. But no one will ever understand how the joint causation of experiential properties by outwardly observable properties and remote properties works. Nor, perhaps, will people ever fully understand various other aspects of human beings that arguably involve remote properties, such as the format in which memories are stored or the ultimate sources of some hard-to-explain behavior.

This may be a bitter pill for hard-driving intellects like Francis Crick, but let's put the bad news in perspective. Note first that the remote properties of electronuclear fabric are almost certainly not the only aspect of the world that no one will ever understand. It is extremely unlikely, for example, that human understanding will ever reach beyond the vast expanse that ballooned out of the Big Bang, let alone grasp the full scope of existence. So it seems that we must make our peace with a certain amount of permanent mystery, whatever the prospects are for understanding how experience is generated. Second, I question whether a backdrop of permanent mystery is such a bad thing. If religious people can feel comfortable believing that they cannot understand the ways of God, then I think all people should be able to feel comfortable believing that there are aspects of the natural world, including aspects of ourselves, that human beings can never fully understand. One can even make a case that a backdrop of permanent mystery is good for the (electronuclear) soul. I think there is something uplifting, even inspirational, in the thought that complete understanding is not possible. The world commands more respect, more reverence, for being more than anyone can ever know.

One lives and then dies in a world that is only partly comprehensible. This is not exactly a cheerful message, but I see no need to be depressed about it. If you find the quadrants view plausible on rational grounds, I believe you can take it to heart without harming your psychological well-being.

Appendix A

The Case against Substance Dualism

This critique of substance dualism supplements chapter 2. When updated to incorporate the modern belief that a living human body is an electronuclear system, substance dualism is the view that a living human being consists of an electronuclear body plus a mind or soul object that is made of nonelectronuclear stuff.

Some writers equate substance dualism with the dualism of Descartes, but in fact substance dualism is a broader category that includes the dualism of Descartes as its most famous example. To evaluate substance dualism fairly, one should consider the full range of possible varieties. These can differ from one another in the following five respects.

1) Lifespans of nonbodily souls. Most and maybe all advocates of substance dualism believe that a person's soul separates from the body when the body dies, and continues to live thereafter and indeed forever. There is less agreement about when a person's soul begins its existence. In any case, no belief about the lifespan of the soul is a logical consequence of the core idea of substance dualism. The belief that a human being consists of a body made of electronuclear fabric and a soul made of something else can be combined with various beliefs about the lifespan of the soul.

2) Possessors of nonbodily souls. Descartes held that every human being has a nonbodily soul, but no nonhuman animal does. Many substance dualists share this belief. However, there are substance dualists who hold that some nonhuman animals have nonbodily souls too. If substance dualism is correct, there must be a line that separates the living beings that have nonbodily souls from the living beings that are exclusively electronuclear systems. But substance dualists can disagree about where this line runs.

3) *Geometry of nonbodily souls.* Descartes held that bodies are "extended things," while nonbodily souls are "unextended things." His claim that bodies are extended things is clear enough, but it is not clear what to make of his claim that nonbodily souls are unextended things. Many writers take it to mean that nonbodily souls are not in space. Here are two typical examples:

> According to Descartes, not only do minds lack spatial extension but also they are not in space at all. (Kim 2006, p.48)

> The mind, according to Descartes, is literally nowhere. It lacks any spatial location. (Jacquette 2009, 16)

I find this interpretation of Descartes questionable for three reasons. First, to my knowledge Descartes never says that souls are not in space; he simply says again and again that they are "unextended things." Perhaps writers who interpret Descartes in this way are reacting to statements such as the following in the *Discourse on Method*:

> From this I knew I was a substance whose whole essence or nature is solely to think, and which *does not require any place*, or depend on any material thing, in order to exist. (Descartes 1988, 36; my emphasis)

However, to say that something *does not require any place* in order to exist is different from saying that it is never in a place. One could easily think that a person's soul exists beyond space in an afterlife, but in space while it is joined to a spatially located body in this life. In fact, this is a common view among those who believe in nonbodily souls. It might have been Descartes's view. Second, the claim that there is a part of a living person that is not in space is difficult to make sense of. Accordingly, one might expect a writer who believes this to say something to help his readers grasp his point. Yet Descartes offers no clarification; he just repeats the phrase "unextended thing." Third, Descartes makes certain remarks that seem to give the soul a spatial location while it is joined to a body. For example, in the Sixth Meditation he speculates that the soul communicates with the body exclusively through the pineal gland, and he uses a sailor-and-ship analogy to explain the soul/body architecture:

> Nature also teaches me, by these sensations of pain, hunger, thirst and so on, that I am not merely present in my body as a sailor is present in a ship, but that I am very closely joined and, as it were, intermingled with it, so that I and the body form a unit. (Descartes 1988, 116)

T. H. Huxley gives an alternative interpretation of Descartes in the following passage:

> Descartes has clearly stated what he conceived to be the difference between spirit and matter. Matter is substance which has extension, but does not think; spirit is substance which thinks, but has no extension. It is very hard to form a definite notion of what this phraseology means, when it is taken in connection with the location of the soul in the pineal gland; and I can only represent it to myself as signifying that the soul is a mathematical point, having place but not extension, within the limits of the pineal body. (Huxley 1896a, 189)

It seems to me that Huxley's dimensionless-point-in-space interpretation is as defensible as the more usual not-in-space interpretation, but it has problems too. First, Descartes never says that the soul is a dimensionless point located in space; he simply uses and reuses the phrase "unextended thing." Second, it is not clear that a dimensionless point can be made of a kind of stuff. Something that is made of a kind of stuff and located in space has to have a nonzero size, one might plausibly insist.

In *The Concept of Mind*, Gilbert Ryle seems to interpret Descartes in both of these ways at once without noticing the contradiction. He describes what he calls "the official doctrine, which hails chiefly from Descartes," saying that one of its central claims is that "minds are not in space" (Ryle 1949, 11). He then proceeds to name this doctrine "the dogma of the Ghost in the Machine" (Ryle 1949, 15–16). But the machine is in space, so if the ghost is *in* the machine, it would seem to follow that the ghost is also in space. Else what does "in" mean here?

However one interprets Descartes, other views concerning the geometry of nonbodily souls are possible. Most obviously, one can believe in nonbodily souls that are extended in just the way that bodies are extended. According to this view, the body is an extended electronuclear object while the mind or soul is an extended object that is made of some other kind of stuff. Given this premise, there is an unlimited variety of possible specifications regarding the soul's shape, size, and location. One could hold that the soul is roughly spherical and located in the head. One could hold that the soul is roughly hourglass-shaped with its upper portion in the head and its lower portion in the chest. One could hold the view of Henry More, a contemporary of Descartes, that the soul can expand and contract like a balloon or an accordion. More's view is described here by E. A. Burtt:

> For him, spirit too must be extended, though its other qualities are widely different from those of matter. Spirit is freely penetrating, and itself able to penetrate and impart motion to matter; it has absolute powers of contraction and dilation, which means that it can occupy greater or less space at will. "The chief seat of the soul, where she perceives all objects, where she imagines, reasons, and invents, and from whence she commands all the parts of the body, is those purer animal spirits in the fourth ventricle of the brain," but he adds that it is not by any means confined there, it is able to spread throughout the whole body on occasion, and even slightly beyond the limits of the body, as a kind of spiritual effluvium. (Burtt 1954, 136–137)

D. M. Armstrong claims that such views are self-contradictory:

> For two spatial things cannot be at the same place at the same time. To attempt to speak of two spatial things with different properties at the same place at the same time is to speak of just one thing with both sets of properties. (Armstrong 1968, 12–13)

Here Armstrong forgets that part of the usual notion of a nonbodily soul is immunity to collisions with bodies. Ghosts, as commonly imagined, can pass through walls. There are good reasons not to believe in an extended nonbodily soul that occupies some of the same space as the body, but it is a mistake to dismiss this idea as a logical contradiction.

Unscientific surveys that I have conducted suggest that most people who believe in nonbodily souls have not thought much about their size, shape, or location. They don't think that a nonbodily soul is an unextended thing. They don't think that it is roughly spherical, or roughly hourglass-shaped. They don't think that it can expand and contract like a balloon. They don't have any definite conception at all of the geometrical properties of nonbodily souls. The most common position on soul geometry seems to be no position.

4) *Location of interactions between the nonbodily soul and the body.* As already noted, Descartes hypothesized that all interaction between the extended body and the unextended thinking thing takes place in the pineal gland, which has a central location in the brain. He could have designated a different locus of interaction. On the assumption that the nonbodily soul has spatial extension, it could interact with the body at many locations within the region that both occupy.

5) *Capabilities of nonbodily souls.* In the Second Meditation, Descartes describes his nonbodily soul in the following way:

> But what then am I? A thing that thinks. What is that? A thing that doubts, understands, affirms, denies, is willing, is unwilling, and also imagines and has sensory perceptions. (Descartes 1988, 83)

Absent from this list is the long-term storage of beliefs, opinions, personal memories, and vocabulary—things that are not types of thinking but that are closely related to thinking. It seems that Descartes confined the capabilities of the nonbodily soul to processes that a person can pay attention to as they happen, leaving all these long-term storage functions to the brain. This allocation of capabilities gives rise to the following puzzle. When the soul separates from the body at death, does it leave behind in the decaying brain all of the person's accumulated beliefs, opinions, memories, and vocabulary, or is all of this stored content "uploaded" from the brain to the departing soul? The first choice leaves the surviving soul psychologically impoverished and cut off from the earthly life that it has just completed. The second choice involves a weird operation and makes one wonder why the brain was ever used for storage in the first place. An alternative that avoids this puzzle (but gives rise to other puzzles) is to assign all these capabilities, including the long-term storage functions, to the nonbodily soul.

<>

Following are five reasons why I find substance dualism extremely implausible. They apply equally to all the varieties of substance dualism that I have just surveyed.

First, scientists have found no signs of soul-to-body influence. If the body is sometimes influenced by a nonbodily soul—for example, when an idea for action leads to action—then one might think that scientists could find telltale ripples in the body at the point or points where the nonbodily influences impinge. Nothing of the sort has been found. This argument is inconclusive because a nonbodily soul could conceivably affect the body in an extremely subtle way, making the ripples difficult to detect. Yet it does have force, because the longer scientists go without discovering a place where a nonbodily soul affects the body, the stronger the suspicion becomes that there is no such place.

Second, scientists have found signs of brain involvement in all the activities that Descartes assigned to the unextended thinking thing. All of these activities can be adversely affected by specific types of brain disease, brain injury, or brain surgery. Numerous examples have been recorded. Modern brain imaging techniques have been used to show that specific regions of the brain are active when a healthy person thinks about a puzzle or has a certain emotional feeling. The accumulation of such evidence shows that the electronuclear brain is heavily involved in all the activities that Descartes and others have assigned to a nonbodily soul. It is true that this leaves open the possibility that a nonbodily soul also plays a role in these activities, partnering with the brain in some unspecified way. But with the brain so heavily involved, it is not clear that there is a significant role left for a nonbodily soul.

Third, there is no credible account of how a body and a nonbodily soul become joined. An egg is fertilized, a fetus grows in the womb, and a baby is born. If substance dualism is correct, then somewhere along the way the person's nonbodily soul must link up with the electronuclear embryo. Where does the nonbodily soul come from and how does it link up with the growing electronuclear system?

Fourth, there is no credible account of how the process of evolving electronuclear systems teamed up with nonbodily souls. Over a period of three or four billion years, earth's evolutionary process has successively brought many life forms into existence, including bacteria, protozoa, invertebrates, vertebrates, primates, and humans. If substance dualism is correct, then somewhere along the way the first individual with a non-bodily soul was born. Where did that first nonbodily soul come from and how did it get into the evolutionary process? Of course, this worry does not arise for those who do not believe in evolution, but denying evolution comes at the price of massive evidence-denial.

Fifth, there is no credible way to draw a line between creatures that have nonbodily souls and creatures that do not have nonbodily souls. Wherever this line runs, there are likely to be creatures without nonbodily souls that have capabilities similar to those that belong to the nonbodily souls of other creatures. This forces a substance dualist to say that these capabilities are located in the nonbodily souls of one species but in the electronuclear bodies of another species. For example, a substance dualist who believes that only human beings have nonbodily souls seems obliged to say that nonbodily souls play an important role in human vision, but no role at all in chimpanzee vision. Such a sharp cutover from a purely electronuclear implementation of a capability to a partially soul-based implementation of the same capability defies explanation. Descartes dispensed with this problem by declaring that all nonhuman animals are automatons without any sensations, thoughts, or feelings, but the case against that claim is very strong.

A respectable defense of substance dualism would have to include plausible responses to all five of these criticisms.

<>

Descartes made several arguments for his version of substance dualism. All of these arguments are marked by logical gaps or oversights, as explained below.

Doubting. In his Second Meditation, Descartes claims that he can doubt that his body exists but he cannot doubt that his thoughts exist, and from this he concludes that his thoughts are in a separate mind object that is linked to his body object. One issue here is exactly what kind of mental gyrations Descartes put himself through in the name of doubting. In trying to replicate Descartes's procedure, I have never succeeded in doubting the existence of my body. Moreover, at the end of his Sixth Meditation, Descartes states that "the exaggerated doubts of the last few days should be dismissed as laughable" (1988, 122). This statement suggests that he never had a genuine doubt about the existence of his body, but only a pseudo-doubt cooked up for rhetorical purposes. Let's suppose, though, that Descartes really did manage to doubt the existence of his body, or at least that he had a thought about his body that was somehow doubt-like. This would not support an inference to substance dualism, because it is compatible with a body-only view. If a person is a complex electronuclear system, then it is possible that his cognitive perspective on his hands differs from his cognitive perspective on his thoughts in a way that enables him to doubt the existence of one but not the other.

Divisibility. In the Sixth Meditation, Descartes argues that the mind and the body must be separate things because "the body is by its very nature always divisible, while the mind is utterly indivisible" (1988, 119–120). In support of this claim, he notes that "if a foot or an arm or any other part of the body is cut off, nothing has thereby been taken away from the mind" (1988, 120). This is actually the germ of one of the arguments *against* substance dualism that I presented above. If one applies this amputation test to the brain, as has been done many times through injury, disease, or surgery, then the kinds of capabilities that Descartes assigns to a nonbodily soul *are* taken away. All Descartes's comment shows is that these capabilities are not dependent on a person's feet and arms. Before concluding that "the mind is utterly indivisible," one should try harder to divide it!

Clear understanding. Again in the Sixth Meditation, Descartes writes:

> First, I know that everything which I clearly and distinctly understand is capable of being created by God so as to correspond exactly with my understanding of it. Hence the fact that I can clearly and distinctly understand one thing apart from another is enough to make me certain that the two things are distinct, since they are capable of being separated, at least by God. (Descartes 1988, 114)

It is ironic that an appeal to clear understanding should itself be so unclear. Is Descartes claiming that whatever a person can "clearly and distinctly understand" is in fact the case? In addition to being baseless, this claim is self-refuting. One can often clearly understand two mutually exclusive hypotheses, which cannot both be the case. I can understand the idea of an unextended thinking thing linked to an extended unthinking thing. I can also understand the idea of a complex electronuclear system that thinks. Simply contemplating ideas in this manner, no matter how clearly, tells us nothing about how things actually are.

Human beings are not machines. In Part V of the *Discourse on Method* (1988, 43–46), Descartes argues that human beings are not machines because the variety and flexibility of human speech and behavior far exceed what machines are capable of. He then makes the additional claim that the source of the human flexibility that is impossible for machines is the nonbodily soul. This argument suffers from the lack of a definition of the key term "machine." It faces the following dilemma. If "machine" is understood narrowly as the kind of man-made device that existed in Descartes's day, then it seems clear that human beings are not machines. However, in this case it also seems very likely that there are electronuclear systems that are not machines, so the possibility that human beings are electronuclear systems remains open. Combining a clunky seventeenth-century machine and a nonbodily soul is not the only possibility. On the other hand, if "machine" is understood so broadly that every electronuclear system is a machine, there is no longer any reason to think that human beings are not machines. So again, the possibility that human beings are electronuclear systems remains open.

<>

In sum, there are many reasons to consider substance dualism implausible, and there is nothing in Descartes's arguments for substance dualism that effectively counters these reasons.

Appendix B

Unusual Accounts of Outward Observation

This appendix supplements chapter 3. It discusses four challenges to the claim that we observe objects by means of spatial transmissions that travel from the observed objects to our sense organs. These challenges were championed, respectively, by Descartes, Berkeley, Kant, and Hume.

Descartes's version of substance dualism is discussed in appendix A. An alternative to substance dualism that he imagined and rejected is that we are all dupes of a massive deception foisted on us by a powerful evil spirit; none of the objects that we think we see, including our own bodies, actually exist. In thinking that vision informs us about objects from which light travels to our eyes, we are mistaken. At the end of his First Meditation, he writes:

> I will suppose therefore that not God, who is supremely good and the source of truth, but rather some malicious demon of the utmost power and cunning has employed all his energies in order to deceive me. I shall think that the sky, the air, the earth, colours, shapes, sounds and all external things are merely delusions of dreams which he has devised to ensnare my judgement. I shall consider myself as not having hands or

eyes, or flesh, or blood or senses, but as falsely believing that
I have all these things. (Descartes 1988, 79)

Descartes says nothing about *how* such an evil spirit might produce human visual experience, but it is obviously not by directing deceptive streams of light into people's eyes, because by the hypothesis Descartes lays out here people have no eyes. He imagines visual experience to result from a transaction of an unspecified sort between an evil spirit and a nonbodily "thinking thing."

Descartes rejects this hypothesis. I think he is right to do so, of course, but I fault the reason that he gives for this rejection. He presents an alleged proof of the existence of God, and then claims that God would not permit a person to be so thoroughly deceived. Claiming that God has given him "a very great inclination to believe that these ideas come from corporeal objects," Descartes reasons as follows:

So I do not see how God could be understood to be anything
but a deceiver if the ideas were transmitted from a source
other than corporeal things. It follows that corporeal things
exist. (Descartes 1988, 116)

Alleged proofs of the existence of God are notoriously controversial, and the one offered by Descartes is no exception. Hopefully there is a better reason than this to believe that we live among objects from which light travels to our eyes.

There is a better reason, and it is this. We have an excellent general method for assessing the relative credibility of any set of competing theories. In a nutshell, this method consists of checking each theory for clarity and logical consistency and then looking for evidence that favors one clear and logically consistent theory over the others. Applying this method to the choice between the spatial-transmissions-from-objects theory and the evil-spirit theory yields the following result.

The spatial-transmissions-from-objects theory seems logically consistent and reasonably clear, so we can proceed directly to supporting evidence. There is a lot of it, including the following three points. First, cameras that are designed to be sensitive only to light can produce images that closely resemble the original object. This shows that light has the ability to encode much of the content of visual experience. Second, study of the eye and brain reveals an elaborate light-sensitive system. Third, outward observation is marked by an extensive set of regularities. One sees basically the same objects from day to day. In many cases, one can touch, hear, smell, and/or taste the same object that one sees. Several people standing next to one another see basically the same objects, as they can verify by verbally comparing notes. All of these regularities have a straightforward explanation if there are objects radiating spatial transmissions that can be intercepted by the sense organs of any individual.

The evil-spirit theory also seems logically consistent, but it is not especially clear because Descartes describes it so sketchily. The evil spirit's motive and means are not specified, and these details are not easy to fill in. It is difficult to specify a plausible motive because deceiving people in a way that they generally find agreeable, without ever dropping a hint that it's a deception, seems more weird than evil. Perhaps Descartes should have called this being a weird spirit. It is difficult to specify a plausible means because everything that could function as a means has been hypothesized not to exist. All we have is an evil spirit and some nonbodily thinking things; one can scarcely even think about such a minimal arrangement.

Moving now to the question of supporting evidence, the situation is simple: there is none. There is no evidence for the existence of such an evil spirit, and no evidence for any particular motive or means. This is not because supporting evidence is inconceivable. One can easily imagine experiences that would support the evil-spirit theory. Some seeming object could lose its usual stability and start to fade or flicker in a peculiar way. Or your vision could indicate an object where your hands encounter nothing. Such anomalies would suggest a technical malfunction that could be explained by the existence of a hidden technician. Nothing of the sort ever happens. One can also imagine hearing a cackling laugh followed by an eerie voice that says "This is the evil spirit speaking. With the exception of this important announcement, every aspect of your experience is a deception that I am foisting on you. Your life is a fraud. Have a nice day." Humanity has received no such message.

In sum, the spatial-transmissions-from-objects theory is clear, logically consistent, and supported by considerable evidence, whereas the evil-spirit theory, though logically consistent, is not very clear and is not supported by any evidence. Accordingly, we have much more reason to believe the spatial-transmissions-from-objects theory. But, one might ask, couldn't all this supporting evidence for the spatial-transmissions-from-objects theory be part of the evil spirit's elaborate deception? Yes, this is possible. The tests of clarity and evidence do not absolutely rule out the evil-spirit theory; they only show that we have much more reason to believe the alternative theory that we have bodies with eyes that receive spatial transmissions from objects. The evil-spirit theory loses the contest of reasons.

<>

A different alternative to the spatial-transmissions-from-objects theory was described, and also defended, by Berkeley. According to him, the world consists exclusively of spirits that perceive and think. There are no objects other than spirits. Moreover, there is no space that the spirits are in: "all extension exists only in the mind" (Berkeley 1988, 78). The spirits exist spacelessly, neither close together nor far apart. Each human being is a spirit. There is also a spirit with extraordinary powers, commonly called God. Whether Berkeley thinks that nonhuman animals are spirits or merely perceptions in the minds of human spirits and God is not clear to me; a noteworthy characteristic of his work is that he rarely mentions nonhuman animals. There is a passage in which he says that "the visive sense" was "bestowed on animals" with the apparent purpose of helping them cope with their environment (Berkeley 1963, 38, paragraph LIX). There is another passage in which he compares "the visive faculty" of mites with that of men (Berkeley 1963, 51, paragraphs LXXX and LXXXI). These passages suggest that he thinks of many animals, even lowly mites, as perceiving spirits. Yet he never claims this explicitly, and one has to wonder because such a view would seem to be at odds with his Christian faith.

As for the source of human visual experience and human perceptual experience in general, Berkeley's theory is that God has a vast amount of perceptual experience and "imprints" or "impresses" or "excites" each human spirit with selected portions of it. Human perceptual experience thus comes from God's perceptual experience by a kind of copying or sharing process. Although motion picture technology did not exist at the time that Berkeley wrote, it seems to me that he was in effect thinking of the world apart from the spirits that perceive it as a kind of movie running in the mind of God.

An important feature of this account, which Berkeley mentions but does not emphasize, is that God's "perception" is not the same thing as human perception. Human perception gives people access to portions of a sensory field that already exists in God's mind, whereas God's "perception" creates this sensory field. In its creative aspect, God's "perception" is more like human imagination or dreaming. In human terms, it is a kind of hybrid power—perceptually rich imagination or imagination-based perception. It is a power that people do not possess, which is similar to human perception in some ways but different in others.

Berkeley defends his theory in a surprising manner. Rather than arguing that it is favored by the available evidence, he claims that the alternative, overwhelmingly popular spatial-transmissions-from-objects theory can be ruled out because it contains a logical contradiction. He defends this claim in several passages, of which the following is typical:

> It is indeed an opinion strangely prevailing amongst men, that houses, mountains, rivers, and in a word all sensible objects have an existence natural or real, distinct from their being perceived by the understanding. But with however great an assurance and acquiescence this principle may be entertained in the world; yet whoever shall find in his heart to call it in question, may, if I mistake not, perceive it to involve a manifest contradiction. For what are the forementioned objects but the things we perceive by sense, and *what do we perceive by sense besides our own ideas or sensations*; and is it not plainly repugnant that any one of these or any combination of them should exist unperceived? (Berkeley 1988, 54, paragraph 4; my emphasis)

The trouble with this argument is that it overlooks an important ambiguity of the word "perceive." People who believe that objects exist independently of being perceived sometimes speak of perceiving those objects, and sometimes of perceiving their own sensations. Thus, there is a simple response to the italicized question in the quoted passage: we also perceive independently existing objects, in the sense that we have cognitive encounters with them that are mediated by perceptual experience. One can say that one perceives one's own perceptual experience directly and independently existing objects indirectly, by means of one's perceptual experience. Given a proper appreciation of the different ways in which the word "perceive" is used, Berkeley's brash attack on the ordinary view collapses.

There are strong arguments against Berkeley's theory. To begin with, it contains some apparent contradictions, or at a minimum some very ragged edges. One ragged edge concerns human behavior. Berkeley accepts the common belief that people do things that have effects on what they and others can perceive. He considers what happens "when I excite a motion in some part of my body" (Berkeley 1988, 96, paragraph 116). He describes the activity of one person observing other people's behavior in the following terms:

> From what has been said, it is plain that we cannot know the existence of other spirits, otherwise than by their operations, or the ideas by them excited in us. I perceive several motions, changes, and combinations of ideas, that inform me there are certain particular agents like myself, which accompany them, and concur in their production. (Berkeley 1988, 108, paragraph 145)

This threatens to contradict his claim that all our perceptual experience is excited in us *by* God. Berkeley's solution seems to be to carve out an exception to that general claim, dividing human perceptions into those that come from God and those that come from the actions of human spirits:

> Lastly, I have nowhere said that God is the only agent who produces all the motions in bodies. It is true, I have denied there are any other agents besides spirits: but this is very consistent with allowing to thinking rational beings, in the production of motions, the use of limited powers, ultimately indeed derived from God, but immediately under the direction of their own wills, which is sufficient to entitle them to all the guilt of their actions. (Berkeley 1988, 184)

> But though there be some things which convince us, human agents are concerned in producing them; yet it is evident to everyone, that those things which are called the works of nature, that is, the far greater part of the ideas or sensations perceived by us, are not produced by, or dependent on the wills of men. There is therefore some other spirit that causes them, since it is repugnant that they should subsist by themselves. (Berkeley 1988, 108, paragraph 146)

This is unsatisfactory, for two reasons. First, one cannot draw a clean line between perceptions that are produced by "human agents" and perceptions of "works of nature." How should one classify perception of a field of crops planted by people, perception of a sky made smoky by a man-made fire, or perception of a lake kept in existence by a man-made dam? Are these perceptions of "works of nature" excited in us by God or perceptions of the results of human action? In a similar vein, what does Berkeley's theory say about my perceptions of a person who is motionlessly sleeping or sunning himself on a beach? Does it make sense that these are perceptions from God, but if the person stirs then my perceptions of the movement are the result of human action? Second and more fundamentally, it is obscure how human action can produce perceptual experience at all, according to Berkeley's theory. God is held to have special powers that enable Him to make parts of His self-created perceptual experience available to human spirits through a copying or sharing process. But how does one person's intentional action of taking a walk give someone else the perceptual experience of a person walking? I see nothing in Berkeley's theory that can explain this.

Could Berkeley's theory be improved by eliminating this exception for human action? Suppose he had said that human action never produces any perceivable effects; it only *seems* to produce perceivable effects because God has perceptions of human action that harmonize perfectly with what human agents try to do. I don't see this as an improvement because it creates a strong case that God is deceiving us, which is incompatible with God's benevolence. If my unshakeable impression that I can move my body and produce perceivable effects is completely mistaken, and if this arrangement was established by God, then God is deceiving me.

A second ragged edge of Berkeley's theory concerns the role of sense organs in producing human perceptual experience. Scientists study the way light is refracted by the lens of the eye, strikes the retina, and triggers activity in nerve cells. This process is almost universally believed to play a key role in producing visual experience. Moreover, Berkeley himself wrote a book, *Essay towards a New Theory of Vision*, in which he seems to accept this. It describes the paths of light rays and even includes diagrams that show various paths that light rays can take through the eye. How can this kind of account of the processing of light by the eyes be reconciled with the claim that "all extension exists only in the mind" or the claim that visual experience is excited in us by God? The only reconciliation that I can think of is that the process involving light rays and eyes does not actually help produce visual experience. Rather, it is part of the stream of perceptions in God's mind, at the end of which God excites visual experience in the human spirit that is associated with the merely perceived "eyes." This creates another strong case that God is deceiving us. If my eyes play no role in producing my visual experience despite all the evidence that they do, and if this arrangement was established by God, then God is deceiving me.

More fundamentally, it is hard to understand why people even have eyes and other sense organs if human perceptual experience is simply a copy of some of God's perceptual experience. Copying perceptual experience from one spirit to another would seem to be a job for mental telepathy of some sort. Sense organs, to all appearances, are transducers—devices that play a role in generating perceptual experience from something else. If this is not what sense organs are, and if our perceptions of sense organs are delivered to us by God, then there is a strong case that God is deceiving us.

If there is a way to remove these ragged edges from Berkeley's theory (a very big "if"), the next step in evaluating the theory is the test of evidence. Here, Berkeley's perceptual-experience-from-God theory is in the same boat as the perceptual-experience-from-an-evil-spirit theory rejected by Descartes. According to both of these theories, all of our perceptual experience is delivered to us by a hidden master technician. But there is no evidence for the existence of either of these hidden master technicians or for any process by which a hidden master technician could deliver perceptual experience to us. Some people think there is evidence for the existence of God—perhaps in the form of natural wonders, miracles, or ancient texts—but I suspect that even these people will agree that there is no evidence that God delivers our moment-to-moment perceptual experience to us in the way that Berkeley claims. The usual view of people who believe in God is that God created a world of just the sort that Berkeley says does not exist—a world of objects in space that human beings can outwardly observe by means of light, sound, and other spatial transmissions.

<>

Kant describes and defends an alternative to the spatial-transmissions-from-objects theory that is characterized by what he calls the "transcendental ideality" of space and time. He maintains that space and time "belong merely to the subjective constitution of our manner of sensibility" (Kant 1965, 73) and "are merely subjective conditions of all our intuition" (Kant 1965, 86). He says that "if the subject, or even only the subjective constitution of the senses in general, be removed, the whole constitution and all the relations of objects in space and time, nay space and time themselves, would vanish" (Kant 1965, 82). This much accords with Berkeley's statement that "all extension exists only in the mind." However, Kant does not share Berkeley's view that reality consists exclusively of spirits or that human perception comes from God by way of a spirit-to-spirit copying process. Rather, Kant says there are objects, which he calls "things-in-themselves," that are responsible for the object-like portions of visual and other perceptual experience. Since space and time are purely mental, these things-in-themselves do not send spatial transmissions to our sense organs. Their connection to the object-like portions of perceptual experience is of a different kind. Of what kind exactly is a matter of great obscurity, however. Kant says that we are "affected by objects" (1965, 71, 78, 93). He also says that objects are "given to us" (1965, 92), "given to our senses" (1965, 78), and "given to us through the senses" (1965, 78). But he does not explain how all this affecting and giving can take place outside space and time. In addition, he says that the object-like portions of perceptual experience do not enable us to learn anything about the things-in-themselves that are somehow responsible for them. Whatever the connection between things-in-themselves and perceptual experience is, it does not include a bridge for cognition. The nature of the nonspatial and nontemporal things-in-themselves is completely unknowable.

If you are like me in being unable to imagine the situation that Kant is describing here, this is a sufficient reason to reject it. He is certainly right that perceptual experience has a spatial and a temporal aspect, but the claim that there is no space and time apart from perceptual experience is impossible to make sense of. How can the things-in-themselves that exist independently of everyone's experience not be in a space and a time that also exist independently of everyone's experience?

In addition to being so strange, Kant's theory seems unable to answer some very basic questions. First, what is the difference, if any, between the space that one sees in a mirror and the space that the mirror is in? It is hard to see how one can draw a distinction here using only Kant's notion of mental space. But it is also hard to see how one can give a plausible description of the world that does not draw a distinction here. Second, what are we to make of the ordinary scientific study of perceptual processes, such as the way light travels through space to the retina, or the way compression waves travel through space to the eardrum? If scientists who think they are studying the way spatial transmissions from objects impinge on our sense organs are not doing that, what are they doing? How can Kant avoid the drastic conclusion that all this scientific study of perceptual processes is a sham? It seems to me that the only way out of these difficulties is to acknowledge a distinction between the space that is an aspect of perceptual experience and the experience-independent space that is occupied by electronuclear objects.

Kant's main argument for the "transcendental ideality" of space is that human beings can know the theorems of geometry with a kind of certainty that would not be possible if space were not purely mental. The trouble with this argument is that it presupposes that there is only one kind of space. The argument melts away as soon as one distinguishes different kinds of space—the space that is an aspect of visual experience, the space that is occupied by electronuclear objects, the spaces that a mathematician can imagine, and perhaps more. Drawing these distinctions enables one to say that the kind of certainty one can have regarding a theorem of geometry concerns a space that one is merely imagining. Whether that theorem also applies to the space that is occupied by electronuclear objects is not so certain. This point became glaringly obvious in the nineteenth century when mathematicians realized that they were able to imagine different spaces—not just Euclidean but also hyperbolic and elliptic—and to prove theorems that were true in one of these spaces but not in the other two. Which of these theorems describe the space that is occupied by electronuclear objects was seen to be a matter of great uncertainty.

<>

Hume does not advocate an alternative theory. Rather, his position is that the spatial-transmissions-from-objects theory might be true, but we have no reason to believe it. He justifies his agnostic neutrality as follows:

> It is a question of fact, whether the perception of the senses be produced by external objects, resembling them: how shall this question be determined? By experience surely; as all other questions of a like nature. But here experience is, and must be entirely silent. The mind has never anything present to it but the perceptions, and cannot possibly reach any experience of their connexion with objects. The supposition of such a connexion is, therefore, without any foundation in reasoning. (Hume 2004, 119)

This seductive passage overlooks two relevant points. First, scientists can and do study the spatial transmissions and neural processes that connect objects to our perceptions of them, so it is simply not the case that we "cannot possibly reach any experience" of these connections. Second, no one has ever succeeded in describing an even vaguely plausible alternative to the spatial-transmissions-from-objects theory. As I have explained, the evil-spirit theory rejected by Descartes, the perceptions-from-God theory defended by Berkeley, and the "transcendental ideality" of space theory defended by Kant all have huge problems. The spatial-transmissions-from-objects-theory cannot be wrong unless some other theory is right. What could that other theory be? Hume is thus mistaken when he claims that the spatial-transmissions-from-objects theory is "without any foundation in reasoning." There is formidable reasoning that supports it.

Hume goes on to recommend that people should live as if the spatial-transmissions-from-objects theory is true, even though they have no reason to believe it. He calls this peculiar state of affairs

> ... the whimsical condition of mankind, who must act and reason and believe; though they are not able, by their most diligent inquiry, to satisfy themselves concerning the foundations of these operations, or to remove the objections, which may be raised against them. (Hume 2004, 125)

This is not the situation we are in. The condition of mankind may be whimsical in some respects, but not in this one. The spatial-transmissions-from-objects theory is as well-founded as any specific finding of modern science. Hume simply ignores the considerations that support it.

References

Abbott, Edwin A. 2005. *Flatland: A Romance of Many Dimensions*. New York: New American Library.

Armstrong, D. M. 1968. *A Materialist Theory of the Mind*. New York: Humanities Press.

Balcombe, Jonathan. 2006. *Pleasurable Kingdom: Animals and the Nature of Feeling Good*. New York: MacMillan.

Bardon, Adrian. 2013. *A Brief History of the Philosophy of Time*. New York: Oxford University Press.

Berkeley, George. 1963. *A New Theory of Vision and other Writings*. London: J. M. Dent & Sons.

————. 1988. *Principles of Human Knowledge and Three Dialogues between Hylas and Philonous*. London: Penguin Books.

Bickerton, Derek. 2009. *Adam's Tongue: How Humans Made Language, How Language Made Humans*. New York: Farrar, Straus and Giroux.

Borst, C. V., ed. 1970. *The Mind-Brain Identity Theory*. London: MacMillan.

Broad, C. D. 1925. *The Mind and Its Place in Nature*. New York: Harcourt, Brace.

Brooks, Michael. 2008. *13 Things That Don't Make Sense: The Most Baffling Scientific Mysteries of Our Time*. New York: Doubleday.

Burtt, Edwin Arthur. 1954. *The Metaphysical Foundations of Modern Science.* Garden City, NY: Doubleday.

Campbell, Keith. 1984. *Body and Mind.* Notre Dame, IN: University of Notre Dame Press.

Campbell, Neil. 2003. *Mental Causation and the Metaphysics of Mind: A Reader.* Peterborough, Ontario: Broadview Press.

Cave, Stephen. 2012. *Immortality.* New York: Crown Publishers.

Chalmers, David J. 1996. *The Conscious Mind: In Search of a Fundamental Theory.* New York: Oxford University Press.

———. 2010. *The Character of Consciousness.* New York: Oxford University Press.

Churchland, Paul M. 1988. *Matter and Consciousness: A Contemporary Introduction to the Philosophy of Mind.* Cambridge, MA: MIT Press.

———. 1995. *The Engine of Reason, the Seat of the Soul: A Philosophical Journey into the Brain.* Cambridge, MA: MIT Press.

Cobb, John B., Jr. and David Ray Griffin, eds. 1978. *Mind in Nature: Essays on the Interface of Science and Philosophy.* University Press of America.

Crick, Francis. 1995. *The Astonishing Hypothesis: The Scientific Search for the Soul.* New York: Simon & Schuster.

Davidson, Donald. 2001. *Essays on Actions and Events.* Oxford, UK: Oxford University Press.

———. 2001a. "The Individuation of Events." In Davidson 2001.

———. 2001b. "Mental Events." In Davidson 2001.

———. 2001c. "Psychology as Philosophy." In Davidson 2001.

De Caro, Mario and David Macarthur. 2008. *Naturalism in Question*. Cambridge, MA: Harvard University Press.

De Chardin, Pierre Teilhard. 1959. *The Phenomenon of Man*. New York: Harper & Brothers.

Dehaene, Stanislas. 2014. *Consciousness and the Brain: Deciphering How the Brain Codes Our Thoughts*. New York: Viking.

Dennett, Daniel C. 1987. *The Intentional Stance*. Cambridge, MA: MIT Press.

———. 1991. *Consciousness Explained*. Boston: Little, Brown.

———. 2003. "Who's On First? Heterophenomenology Explained." *Journal of Consciousness Studies*, Vol. 10, No. 9–10 (September/October).

———. 2005. *Sweet Dreams: Philosophical Obstacles to a Science of Consciousness*. Cambridge, MA: MIT Press.

De Quincey, Christian. 2002. *Radical Nature: Rediscovering the Soul of Matter*. Montpelier, VT: Invisible Cities Press.

Descartes, Rene. 1988. *Descartes: Selected Philosophical Writings*. Cambridge, UK: Cambridge University Press.

Dretske, Fred. 1995. *Naturalizing the Mind*. Cambridge, MA: MIT Press.

Eddington, Arthur. 1958. *The Nature of the Physical World*. Ann Arbor: University of Michigan Press.

Edelman, Gerald M. 1989. *The Remembered Present: A Biological Theory of Consciousness*. New York: Basic Books.

———. 1992. *Bright Air, Brilliant Fire: On the Matter of the Mind*. New York: Basic Books.

———. 2004. *Wider than the Sky: The Phenomenal Gift of Consciousness*. London: Penguin Group.

Edelman, Gerald M. and Giulio Tononi. 2000. *A Universe of Consciousness: How Matter Becomes Imagination*. New York: Basic Books.

Ells, Peter. 2011. *Panpsychism: The Philosophy of the Sensuous Cosmos.* Winchester, UK: O-Books.

Feigl, Herbert. 1967. *The "Mental" and the "Physical": The Essay and a Postscript.* Minneapolis: University of Minnesota Press.

Ferguson, Niall. 2008. *The Ascent of Money: A Financial History of the World.* New York: Penguin Press.

Flanagan, Owen. 1991. *The Science of the Mind, 2nd ed.* Cambridge, MA: MIT Press.

————. 1992. *Consciousness Reconsidered.* Cambridge, MA: MIT Press.

Fodor, Jerry A. 1983. *The Modularity of Mind: An Essay in Faculty Psychology.* Cambridge, MA: MIT Press.

Franken, Dirk, Attila Karakus and Jan G. Michel, eds. 2010. *John R. Searle: Thinking about the Real World.* Frankfurt: ontos verlag.

Gillett, Carl and Barry Loewer, eds. 2001. *Physicalism and Its Discontents.* Cambridge, UK: Cambridge University Press.

Gleick, James. 2011. *The Information: A History, A Theory, A Flood.* New York: Pantheon Books.

Grandin, Temple and Catherine Johnson. 2005. *Animals in Translation: Using the Mysteries of Autism to Decode Animal Behavior.* New York: Scribner.

Gray, Jeffrey. 2004. *Consciousness: Creeping Up on the Hard Problem.* New York: Oxford University Press.

Griffin, David Ray. 1998. *Unsnarling the World-Knot: Consciousness, Freedom, and the Mind-Body Problem.* Berkeley: University of California Press.

Griffin, Donald R. 1981. *The Question of Animal Awareness: Evolutionary Continuity of Mental Experience.* New York: Rockefeller University Press.

———. 1992. *Animal Minds.* Chicago: University of Chicago Press.

Guttenplan, Samuel. 1994. "An Essay on Mind." In Guttenplan, Samuel, ed. *A Companion to the Philosophy of Mind.* Oxford, UK: Blackwell.

Haeckel, Ernst. 1901. *The Riddle of the Universe at the Close of the Nineteenth Century.* London: Watts.

Hardin, C. L. 1988. *Color for Philosophers: Unweaving the Rainbow.* Indianapolis: Hackett Publishing.

Hartshorne, Charles. 1968. *Beyond Humanism: Essays in the Philosophy of Nature.* Lincoln, NE: University of Nebraska Press.

———. 1978. "Physics and Psychics: The Place of Mind in Nature." In Cobb and Griffin 1978.

Hill, Christopher S. 1991. *Sensations: A Defense of Type Materialism.* Cambridge, UK: Cambridge University Press.

Honderich, Ted. 1988. *A Theory of Determinism: The Mind, Neuroscience, and Life-Hopes.* New York: Oxford University Press.

———. 2002. *How Free Are You? The Determinism Problem.* New York: Oxford University Press.

———. 2005. *On Determinism and Freedom.* Edinburgh: Edinburgh University Press.

Hume, David. 2004. *An Enquiry Concerning Human Understanding.* New York: Barnes & Noble.

Huxley, Thomas H. 1896. *Method and Results: Essays.* New York: D. Appleton.

———. 1896a. "On Descartes' 'Discourse Touching the Method of Using One's Reason Rightly and of Seeking Scientific Truth.'" In Huxley 1896.

———. 1896b. "On the Hypothesis That Animals Are Automata, and Its History." In Huxley 1896. Reprinted in Campbell 2003.

Huxley, Thomas H. and William Jay Youmans. 1869. *The Elements of Physiology and Hygiene: A Textbook for Educational Institutions*. New York: D. Appleton. Available online at books.google.com.

Jackson, Frank. 1998. *From Metaphysics to Ethics*. New York: Oxford University Press.

———. 2004a. "Epiphenomenal Qualia." In Ludlow 2004.

———. 2004b. "What Mary Didn't Know." In Ludlow 2004.

Jacquette, Dale. 2009. *The Philosophy of Mind: The Metaphysics of Consciousness*. London: Continuum.

James, William. 2007. *The Principles of Psychology*, Vol. 1. New York: Cosimo.

Kahneman, Daniel. 2011. *Thinking, Fast and Slow*. New York: Farrar, Straus and Giroux.

Kant, Immanuel. 1965. *Critique of Pure Reason*. New York: St. Martin's Press.

Kim, Jaegwon. 1993. "The Non-Reductivist's Troubles with Mental Causation." In *Supervenience and Mind: Selected Philosophical Essays*. Cambridge, UK: Cambridge University Press.

———. 1998. *Mind in a Physical World: An Essay on the Mind-Body Problem and Mental Causation*. Cambridge, MA: MIT Press.

———. 2001. "Mental Causation and Consciousness: The Two Mind-Body Problems for the Physicalist." In Gillett and Loewer 2001.

———. 2005. *Physicalism, or Something Near Enough*. Princeton: Princeton University Press.

———. 2006. *Philosophy of Mind*, 2nd ed. Cambridge, MA: Westview Press.

————. 2010. "Two Concepts of Realization, Mental Causation, and Physicalism." In *Essays in the Metaphysics of Mind*. New York: Oxford University Press.

Koch, Christof. 2004. *The Quest for Consciousness: A Neurobiological Approach.* Englewood, CO: Roberts.

Kripke, Saul. 1971. "Identity and Necessity." In Munitz, Milton K., ed. *Identity and Individuation*. New York: New York University Press.

————. 1980. *Naming and Necessity.* Cambridge, MA: Harvard University Press.

Kurthen, Martin. 1995. "On the Prospects of a Naturalistic Theory of Phenomenal Consciousness." In Metzinger, Thomas, ed. *Conscious Experience*. Lawrence, KS: Allen Press/Imprint Academic.

Leibniz, Gottfried. 1971. *The Monadology and other Philosophical Writings.* Oxford, UK: Oxford University Press.

Levine, Joseph. 1983. "Materialism and Qualia: The Explanatory Gap." In Chalmers, David J., ed. 2002. *Philosophy of Mind: Classical and Contemporary Readings*. Oxford, UK: Oxford University Press.

Lucretius. 1995. *On the Nature of Things.* Edited and translated by Anthony M. Esolen. Baltimore: Johns Hopkins University Press.

Ludlow, Peter, Yujin Nagasawa and Daniel Stoljar, eds. 2004. *There's Something About Mary: Essays on Phenomenal Consciousness and Frank Jackson's Knowledge Argument.* Cambridge, MA: MIT Press.

Ludwig, Kirk. 2003. "The Mind-Body Problem: An Overview." In Stich, Stephen P. and Ted A. Warfield, eds. *The Blackwell Guide to Philosophy of Mind*. Oxford, UK: Blackwell.

Lycan, William G. 1996. *Consciousness and Experience.* Cambridge, MA: MIT Press.

Maslin, K. T. 2007. *An Introduction to the Philosophy of Mind, 2nd ed.* Cambridge, UK: Polity Press.

McGinn, Colin. 1982. *The Character of Mind*. Oxford, UK: Oxford University Press.

———. 1991. *The Problem of Consciousness: Essays toward a Resolution*. Oxford, UK: Blackwell.

———. 1993. *Problems in Philosophy: The Limits of Inquiry*. Oxford, UK: Blackwell.

———. 1999. *The Mysterious Flame: Conscious Minds in a Material World*. New York: Basic Books.

———. 2004. *Consciousness and Its Objects*. Oxford, UK: Oxford University Press.

———. 2011. *Basic Structures of Reality: Essays in Meta-Physics*. Oxford, UK: Oxford University Press.

McKinsey, Michael. 1996. "Anti-Individualism and Privileged Access." In Pessin, Andrew and Sanford Goldberg, eds. *The Twin Earth Chronicles: Twenty Years of Reflection on Hilary Putnam's "The Meaning of 'Meaning'"*. Armonk, NY: M. E. Sharpe.

Moore, G. E. 1959. "Proof of an External World." In *Philosophical Papers*. London: George Allen and Unwin.

Mukherjee, Siddhartha. 2010. *The Emperor of All Maladies: A Biography of Cancer*. New York: Scribner.

Nagel, Thomas. 1979. *Mortal Questions*. Cambridge, UK: Cambridge University Press.

———. 1979a. "Panpsychism." In Nagel 1979.

———. 1979b. "Subjective and Objective." In Nagel 1979.

———. 1979c. "What is it like to be a bat?" In Nagel 1979.

———. 1986. *The View from Nowhere*. New York: Oxford University Press.

————. 1995. *Other Minds: Critical Essays 1969-1994.* New York: Oxford University Press.

————. 1995a. "Freud's Anthropomorphism." In Nagel 1995.

————. 1995b. "Freud's Permanent Revolution." In Nagel 1995.

————. 1995c. "Searle: Why We Are Not Computers." In Nagel 1995.

————. 2002 "The Psychophysical Nexus." In *Concealment and Exposure and Other Essays.* New York: Oxford University Press.

————. 2012. *Mind and Cosmos: Why the Materialist Neo-Darwinian Conception of Nature Is Almost Certainly False.* New York: Oxford University Press.

Papineau, David. 2001. "The Rise of Physicalism." In Gillett and Loewer 2001.

————. 2002. *Thinking about Consciousness.* Oxford, UK: Oxford University Press.

Park, Robert. 2000. *Voodoo Science: The Road from Foolishness to Fraud.* New York: Oxford University Press.

Penrose, Roger. 1986. "Big Bangs, Black Holes and 'Time's Arrow.'" In Flood, Raymond and Michael Lockwood, eds. *The Nature of Time.* Oxford, UK: Blackwell.

Piccinini, Gualtiero. 2009. "First-Person Data, Publicity and Self-Measurement." *Philosophers' Imprint*, Vol. 9, No. 9 (October).

Place, U. T. 1956. "Is Consciousness a Brain Process?" In Borst 1970.

Putnam, Hilary. 1981. *Reason, Truth and History.* Cambridge, UK: Cambridge University Press.

Rock, Andrea. 2004. *The Mind at Night: The New Science of How and Why We Dream.* New York: Basic Books.

Rose, Steven. 1992. *The Making of Memory: From Molecules to Mind.* New York: Doubleday.

Rosenberg, Gregg. 2004. *A Place for Consciousness: Probing the Deep Structure of the Natural World*. Oxford, UK: Oxford University Press.

Russell, Bertrand. 1921. *The Analysis of Mind*. London: George Allen and Unwin.

———. 1931. *The Scientific Outlook*. New York: W. W. Norton.

———. 1940. *An Inquiry into Meaning and Truth*. New York: W. W. Norton.

———. 1948. *Human Knowledge: Its Scope and Limits*. New York: Simon and Schuster.

———. 1959. "The Ultimate Constituents of Matter." In *Mysticism and Logic and Other Essays*. London: George Allen and Unwin.

———. 1963. "Reply to Criticisms." In Schilpp, Paul Arthur, ed. *The Philosophy of Bertrand Russell*, Vol. 2. New York: Harper & Row.

———. 1995a. *My Philosophical Development*. London: Routledge.

———. 1995b. *An Outline of Philosophy*. London: Routledge.

———. 1996. "Events, Matter, and Mind." In John G. Slater, ed. *The Collected Papers of Bertrand Russell*. Vol. 10, *A Fresh Look at Empiricism 1927-42*. London: Routledge.

———. 1997. *ABC of Relativity*. London: Routledge.

———. 2004a. "Philosophy in the Twentieth Century." In *Sceptical Essays*. London: Routledge.

———. 2004b. *The Problems of Philosophy*. New York: Barnes & Noble.

———. 2007. *The Analysis of Matter*. Nottingham, UK: Spokesman.

Ryle, Gilbert. 1949. *The Concept of Mind*. New York: Barnes & Noble.

Searle, John R. 1984. *Minds, Brains and Science*. Cambridge MA: Harvard University Press.

———. 1992. *The Rediscovery of the Mind.* Cambridge, MA: MIT Press.

———. 1997. *The Mystery of Consciousness.* New York: New York Review of Books.

———. 1998. *Mind, Language and Society: Philosophy in the Real World.* New York: Basic Books.

———. 2002. *Consciousness and Language.* Cambridge, UK: Cambridge University Press.

———. 2002a. "Animal Minds." In Searle 2002.

———. 2002b. "Consciousness." In Searle 2002.

———. 2002c. "How to Study Consciousness Scientifically." In Searle 2002.

———. 2004. *Mind: A Brief Introduction.* New York: Oxford University Press.

———. 2007. *Freedom and Neurobiology: Reflections on Free Will, Language, and Political Power.* New York: Columbia University Press.

———. 2010a. "The Basic Reality and the Human Reality." In Franken 2010.

———. 2010b. "Reply to 'Subjectivity as the Mark of the Mental.'" In Franken 2010.

Shaffer, Jerome. 1970. "Mental Events and the Brain." In Borst 1970.

Shubin, Neil. 2008. *Your Inner Fish: A Journey into the 3.5-Billion-Year History of the Human Body.* New York: Pantheon Books.

Skrbina, David. 2005. *Panpsychism in the West.* Cambridge, MA: MIT Press.

Slater, John G. 1999. "Russell's Conception of Philosophy." In Irvine, Andrew, ed. *Bertrand Russell: Critical Assessments,* Vol. 3, *Language, Knowledge, and the World.* London: Routledge.

Smart, J. J. C. 1959. "Sensations and Brain Processes." In Borst 1970.

———. 1963a. "Materialism." In Borst 1970.

———. 1963b. *Philosophy and Scientific Realism*. London: Routledge & Kegan Paul.

Stoljar, Daniel. 2004. "Two Conceptions of the Physical." In Ludlow 2004.

———. 2006. *Ignorance and Imagination: The Epistemic Origin of the Problem of Consciousness*. New York: Oxford University Press.

Strawson, Galen. 2008. *Real Materialism and Other Essays*. New York: Oxford University Press.

Strawson, Galen et al. 2006. *Consciousness and Its Place in Nature: Does Physicalism Entail Panpsychism?* edited by Anthony Freeman. Charlottesville, VA: Imprint Academic.

Tallis, Raymond. 2011. *Aping Mankind: Neuromania, Darwinitis and the Misrepresentation of Humanity*. Durham, UK: Acumen.

Taylor, John G. 1999. *The Race for Consciousness*. Cambridge, MA: MIT Press.

Van Gulick, Robert. 2004. "So Many Ways of Saying No to Mary." In Ludlow 2004.

Velmans, Max. 2000. *Understanding Consciousness*. London: Routledge.

———. 2007. "Dualism, Reductionism, and Reflexive Monism." In Velmans, Max and Susan Schneider, eds. *The Blackwell Companion to Consciousness*. Malden, MA: Blackwell.

Wagner, Steven J. and Richard Warner, eds. 1993. *Naturalism: A Critical Appraisal*. Notre Dame, IN: University of Notre Dame Press.

Wright, Sewall. 1978. "Panpsychism and Science." In Cobb and Griffin 1978.

Index

G

gas law 197, 240–241
genes 194–195
genotype and phenotype 90
Gleick, James 271
God 199–201, 332, 342–343, 345–349
Grandin, Temple 32, 65
Gray, Jeffrey 144
Griffin, David Ray 148, 153–154
Griffin, Donald R. 65

H

Haeckel, Ernst 148
hard problem of consciousness ix, 101–103, 142–143
Hartshorne, Charles 148, 154–155
hearing 16, 20, 27–28
Hill, Christopher 290, 302–304
homeopathy 149–150
Honderich, Ted 310, 314–316
Hume 351–352
Huxley, T. H. 10–13, 45, 79, 84, 182–183, 311–313, 335

I

identity theories 62, 70, 80, 277–288, 289–307, 318–324
ignorance hypothesis (Stoljar) 251–268
immortality ix, x, 331
inference from perception (Russell) 232–250
intelligibility 157–158, 190–191, 192–210, 267–268
intentionality (Searle) 169–174
intentions 20, 291, 321, 324–329
interpersonal communication 19–20
intrinsic character (Russell) 222, 232–250, 263, 273–274
introspection 25–26, 186, 295–300
inward access 36–37, 44–45, 50

inward observation 19, 25–41, 42–55, 65–66, 88, 161–181, 185–187, 293–300

J

Jackson, Frank 99–100, 185, 255–256, 257
Jacquette, Dale 13–15
James, William 25, 28, 33, 136
joint causation 110–123, 124–126, 176–177, 207–208, 261, 262, 308–330, 332

K

Kahneman, Daniel 77–79
Kant 349–351
Kim, Jaegwon 8–9, 13–15, 83–84, 88, 103–104, 106, 116–123, 313–318, 324
knowledge argument 185, 255–256, 257
Koch, Christof 58–60, 137–138, 144–145
Kripke, Saul 287–288

L

Leibniz 84–86
lenses 16–17
levels 8–9, 177–180, 326–329
Levine, Joseph 115–117
liberal naturalism (Rosenberg) 91
logical problem of experience (Stoljar) 251–268
lucid dreaming 31
Lucretius 1–6, 17, 213

M

Maslin, Keith 13–15
material objects 6, 83, 87, 96, 188
materialism 80–83, 87, 109, 177

www.ingramcontent.com/pod-product-compliance
Lightning Source LLC
Chambersburg PA
CBHW031817170526
45157CB00001B/87